ORDINARY DIFFERENTIAL EQUATIONS

A SERIES OF PROGRAMMES ON DIFFERENTIAL EQUATIONS

CONSULTANT EDITOR:

A. C. BAJPAI

PROFESSOR OF MATHEMATICAL EDUCATION AND DIRECTOR OF CAMET

GENERAL EDITORS

I. M. CALUS

J. A. FAIRLEY

CAMET

(CENTRE FOR THE ADVANCEMENT OF MATHEMATICAL EDUCATION IN TECHNOLOGY)

DEPARTMENT OF MATHEMATICS

LOUGHBOROUGH UNIVERSITY OF TECHNOLOGY

ORDINARY DIFFERENTIAL EQUATIONS

A PROGRAMMED COURSE FOR STUDENTS OF SCIENCE AND TECHNOLOGY

A. C. BAJPAI
I. M. CALUS
J. HYSLOP

LOUGHBOROUGH UNIVERSITY OF TECHNOLOGY

WILEY - INTERSCIENCE
A DIVISION OF
JOHN WILEY & SONS LTD
LONDON · NEW YORK · SYDNEY · TORONTO

Copyright © 1970 by John Wiley & Sons Ltd., All Rights Reserved. No part of this book may be reproduced stored in a retrieval system, or transmitted in any form or by any means, electronic, mechanical photo copying, recording or otherwise without the prior written permission of the copyright owner

Library of Congress Catalog Card No. 70-135349

ISBN 0 471 04370 2

Printed in England by Page Bros (Norwich) Ltd

EDITORS' PREFACE

This series of programmes on differential equations is designed to provide a set of useful tools for scientists and technologists who are interested in solving those equations which arise in their own work. The emphasis therefore is on developing methods of solution for the types of equation which occur most frequently in physical problems. Consequently, many practical examples are given and the more theoretical aspects, such as existence theorems, have been omitted.

The programmed method of presentation, used throughout, has many advantages. The development of the subject progresses in carefully sequenced steps, with the student proceeding at his own pace. At each stage he has an active part to play by answering a question or solving a problem, and thus he learns by doing. By comparing his own answer with that given in the text, he obtains a continuous assessment of his understanding of the subject up to that point. Explanation of the material covered is given in greater detail than is usually found in standard text-books, especially at points where, in their experience as teachers, the authors find that students often have difficulty.

SI units have been adopted as standard throughout. Thus, for example, 5 A is the abbreviation for 5 amperes and ts for t seconds. However, the standard practice of using italic letters for quantities, e.g. C for capacitance, has not been followed, because italic lettering has been used for the answer frames. Circuit diagrams have been drawn in accordance with BS 3939 Sections 4–7 March 1966 onwards.

In spite of careful checking by the general editors, it is possible that the occasional error has crept through. They would appreciate receiving information about any such mistakes that might be discovered.

A debt of gratitude to the following is acknowledged with pleasure:

Loughborough University of Technology for supporting this venture.

Staff and students of Loughborough University of Technology and other institutions who have participated in the testing of these programmes.

Mrs. June Russell for preparing the camera-ready copy from which the book has been printed.

John Wiley and Sons Ltd for their help and cooperation.

CONTENTS

1. **FIRST ORDER DIFFERENTIAL EQUATIONS**   1:1
2. **SECOND ORDER DIFFERENTIAL EQUATIONS WITH CONSTANT COEFFICIENTS** – Solution by Trial Methods   2:1
3. **DIFFERENTIAL EQUATIONS WITH CONSTANT COEFFICIENTS** – Solution by D-operator Methods   3:1
4. **DIFFERENTIAL EQUATIONS WITH CONSTANT COEFFICIENTS** – Solution by Laplace Transform Methods   4:1
5. **SIMULTANEOUS DIFFERENTIAL EQUATIONS** – Solution by D-operator and Laplace Transform Methods   5:1
6. **LINEAR DIFFERENTIAL EQUATIONS** – Variation of Parameters and Solution in Series   6:1

FIRST ORDER DIFFERENTIAL EQUATIONS

A PROGRAMMED TEXT

A. C. Bajpai
J. Hyslop

INSTRUCTIONS

This programme constitutes a self-instructional course on methods of solving certain first order differential equations.

The programme is divided up into a number of FRAMES which are to be worked *in the order given*. You will be required to participate in many of these frames and in such cases the answers are provided in ANSWER FRAMES, designated by the letter A following the frame number. Steps in the working are given where this is considered helpful. The answer frame is separated from the main frame by a line of asterisks: ******************.
Keep the answers covered until you have written your own response. If your answer is wrong, go back and try to see why. Do not proceed to the next frame until you have corrected any mistakes in your attempt and are satisfied that you understand the contents up to this point.

Before reading this programme it is necessary that you are familiar with the following

Prerequisites

Integration of standard functions. Integration by substitution, by parts and by using partial fractions.

Partial differentiation — first and second partial derivatives and the formula $du = \dfrac{\partial u}{\partial x} dx + \dfrac{\partial u}{\partial y} dy$.

CONTENTS

Instructions

FRAMES

1	Introduction
2 - 16	Definitions and classifications
17	Types of 1st order differential equations
18 - 38	Type I - Variables separable
39 - 49	Type II - Homogeneous
50 - 63	Type III - Linear
64 - 74	Type IV - Exact
75 - 76	Miscellaneous examples
77	Answers to miscellaneous examples

FIRST ORDER DIFFERENTIAL EQUATIONS

FRAME 1

Introduction

In many physical situations, equations arise which involve differential coefficients.

For example, if a body falls freely under gravity we have

$$\frac{d^2s}{dt^2} = g$$

where s is the distance fallen in time t.

Some other examples are:

(i)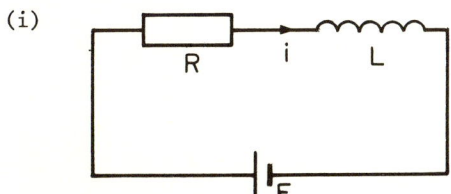

For the circuit shown

$$L\frac{di}{dt} + Ri = E$$

where i is the current at time t.

(ii) When a mass oscillates on the end of a spring and is subject to a frictional resistance proportional to its speed, the equation of motion may be written as

$$m\frac{d^2x}{dt^2} + r\frac{dx}{dt} + sx = 0$$

where x is the displacement from the equilibrium position at time t.

(iii) When a uniform beam is supported at its ends the deflection y at a distance x from one end is given by the equation

$$EI\frac{d^2y}{dx^2} = \frac{1}{2}w(x^2 - \ell x).$$

(iv) Newton's Law of Cooling states that the rate of decrease of temperature of a hot body is proportional to its excess temperature over that of the surroundings. This leads to the equation

$$\frac{d\theta}{dt} = -k(\theta - \theta_o)$$

where θ is the temperature at time t.

In any of these cases the problem is to find the dependent variable in terms of the independent one, e.g. in (ii) we have to find x in terms of t, or in (iii) y in terms of x.

FIRST ORDER DIFFERENTIAL EQUATIONS

FRAME 2

Definitions and Classifications

A relationship between a variable quantity x and a dependent function y and its derivatives $\frac{dy}{dx}$, $\frac{d^2y}{dx^2}$, $\frac{d^3y}{dx^3}$, ... is called an ORDINARY DIFFERENTIAL EQUATION.

Some examples of ordinary differential equations are:

(i) $\quad \frac{dy}{dx} = Kx$

(ii) $\quad x^2(1 + y)\frac{dy}{dx} - (1 + x)y^2 = 0$

(iii) $\quad \frac{d^2y}{dx^2} = -n^2 y$

(iv) $\quad \left(\frac{dy}{dx}\right)^2 + 3y = 0$

(v) $\quad x\frac{d^3y}{dx^3} + \frac{d^2y}{dx^2} + x\left(\frac{dy}{dx}\right)^4 = 0$

FRAME 3

If the highest derivative that occurs in an equation is $\frac{d^n y}{dx^n}$, the equation is said to be of ORDER n.

The DEGREE of a differential equation is the power to which the highest derivative is raised.

Now return to FRAME 2 and write down the order and degree of each differential equation (abbreviated to d.e. in future).

****

3A

	Order	Degree
(i)	1st	1st
(ii)	1st	1st
(iii)	2nd	1st
(iv)	1st	2nd
(v)	3rd	1st

FIRST ORDER DIFFERENTIAL EQUATIONS

FRAME 4

Note that in the last example the degree of the equation is determined by the power of the <u>highest</u> derivative $\frac{d^3y}{dx^3}$, <u>not</u> by the fourth power term in $\frac{dy}{dx}$.

FRAME 5

You will recall from FRAME 2 that an ordinary d.e. was defined as a relationship between a variable quantity x and a dependent function y and its derivatives. As seen in FRAME 1, these equations normally arise from physical situations and it is often required to obtain a functional relationship between x and y alone, having eliminated the derivatives. This relation is referred to as the SOLUTION of the d.e.

FRAME 6

For example, consider the simple d.e. $\frac{dy}{dx}$ = x, which is of the first order and degree.

Straightforward integration yields the <u>solution</u> $y = \frac{1}{2}x^2 + A$, where A is an arbitrary constant of integration.

Verify this by the inverse process of differentiation.

FRAME 7

Now <u>solve</u> the second order d.e. $\frac{d^2y}{dx^2}$ = 0 by two successive integrations.

****

<u>7A</u>

$$\frac{dy}{dx} = A$$
$$\therefore \quad y = Ax + B$$

(A and B are arbitrary constants.)

1:4   FIRST ORDER DIFFERENTIAL EQUATIONS

FRAME 8

From FRAMES 6 and 7 you will note that the solutions of the first and second order d.e.'s considered contained one and two arbitrary constants respectively.

In general, it can be shown that the solution of an ordinary d.e. of order n involves n arbitrary constants.

FRAME 9

Conversely, if a functional relationship between y and x contains n arbitrary constants, the elimination of these constants yields an ordinary d.e. of order n.

We shall consider a case of two arbitrary constants in the relation and show that a second order d.e. is obtained.

Thus if $\quad y = Ae^{2x} + Be^{-x}$, then, on differentiating,

$$y' = 2Ae^{2x} - Be^{-x}$$

and $\quad y'' = 4Ae^{2x} + Be^{-x}$.

From the last two of these equations

$$y' + y'' = 6Ae^{2x}$$

and $\quad y'' - 2y' = 3Be^{-x}$.

Substituting in $\quad 6y = 6Ae^{2x} + 6Be^{-x}$ gives

$$6y = (y' + y'') + 2(y'' - 2y')$$

i.e. $\quad y'' - y' - 2y = 0$.

Alternatively if you are familiar with determinants, the result of the elimination of A and B is given by

$$\begin{vmatrix} y & 1 & 1 \\ y' & 2 & -1 \\ y'' & 4 & 1 \end{vmatrix} = 0$$

FIRST ORDER DIFFERENTIAL EQUATIONS

FRAME 10

Now obtain d.e.'s by eliminating the arbitrary constants from the following relations:

(i) $y = A \cos x$

(ii) $y = A \cos x + B \sin x$

****

10A

(i) $\dfrac{dy}{dx} + y \tan x = 0$

(ii) $\dfrac{d^2y}{dx^2} + y = 0$

FRAME 11

Solutions of the type given in FRAMES 6, 7A and 10 with the appropriate number of arbitrary constants are called GENERAL SOLUTIONS.

FRAME 12

In physical problems solutions are usually required which satisfy certain specified conditions. These conditions provide information from which values may be assigned to the arbitrary constants. This type of solution, which satisfies certain definite conditions, is called a PARTICULAR SOLUTION, and the conditions satisfied are called BOUNDARY CONDITIONS or INITIAL CONDITIONS.

FRAME 13

To illustrate, consider the general solution $y = \tfrac{1}{2}x^2 + A$ to the equation $\dfrac{dy}{dx} = x$ in FRAME 6.

Let us assume that the boundary condition is given to be $y = 1$ when $x = 0$. The value assigned to A in this particular case is obviously $A = 1$ and the particular solution is $y = \tfrac{1}{2}x^2 + 1$.

What is the particular solution to the above equation given that y = 0 when x = 2?

****

13A

$$y = \frac{1}{2}x^2 - 2$$

FRAME 14

How many boundary conditions must be given in the case of a second order d.e. in order that the arbitrary constants may be found?

****

14A

2 (since there are 2 arbitrary constants to be found).

FRAME 15

Now return to the general solution you obtained in FRAME 7A to the second order d.e. $\frac{d^2y}{dx^2} = 0$. Find the particular solution which satisfies the boundary conditions: when x = 1, y = 1 and y' = 2.

****

15A

$$y = 2x - 1,$$

as $1 = A + B$

and $2 = A.$

FIRST ORDER DIFFERENTIAL EQUATIONS

FRAME 16

You now have the background necessary for the study of the solution of ordinary d.e.'s.

You have already solved two trivial d.e.'s by direct integration. In order to solve more difficult d.e.'s additional techniques need to be developed.

Attention will be confined in this programme to <u>d.e.'s of 1st order and 1st degree</u> and furthermore, only those equations will be considered which occur most frequently in science and engineering.

FRAME 17

<u>Types of 1st order d.e.'s</u>

There are four main types of 1st order, 1st degree d.e.'s.

These are:

      I   Variables separable
      II   Homogeneous
      III   Linear
      IV   Exact.

Each of the above types will be discussed under the general headings of:

      (a)   Recognition
      (b)   Techniques of solution
      (c)   Applications.

FRAME 18

<u>Type I - Variables separable</u>

<u>Recognition</u>

Since d.e.'s of the 1st order and 1st degree contain $\frac{dy}{dx}$ to the <u>first</u> power only, they can be written as

$$\frac{dy}{dx} = F(x,y).$$

FIRST ORDER DIFFERENTIAL EQUATIONS

FRAME 18 continued

In many cases F(x,y) may be written as

$$F(x,y) = f(x)\, g(y)$$

where f(x) and g(y) are functions of x and y alone, respectively.
We may then "separate the variables" and write

$$\frac{dy}{g(y)} = f(x)\, dx.$$

Now, in the example

$$\frac{dy}{dx} = x^2 y + y,$$

express the right hand side (R.H.S.) in the form f(x) g(y) and separate the variables.

****

18A

$$\frac{dy}{dx} = (x^2 + 1)y$$

$$\frac{dy}{y} = (x^2 + 1)dx$$

FRAME 19

Now decide which of the following equations are of this type and separate the variables where possible.

(1) $\cos^2 x \dfrac{dy}{dx} = \cos^2 y$

(2) $(1 + x^2)\dfrac{dy}{dx} = 1 + y^2$

(3) $\dfrac{dy}{dx} = \dfrac{x + y}{x}$

(4) $x^2(1 + y)\dfrac{dy}{dx} + (1 - x)y^2 = 0$

(5) $\sin x \cos y \dfrac{dy}{dx} + \cos x \sin y = 0$

(6) $(1 - x)\dfrac{dy}{dx} + (xy + \sin x) = 0$

(7) $xy(1 + x^2)\dfrac{dy}{dx} - y^2 = 1$

****

FIRST ORDER DIFFERENTIAL EQUATIONS

19A

(1) $\dfrac{dy}{\cos^2 y} = \dfrac{dx}{\cos^2 x}$

(2) $\dfrac{dy}{1+y^2} = \dfrac{dx}{1+x^2}$

(3) Not separable

(4) $\dfrac{1+y}{y^2} dy + \dfrac{1-x}{x^2} dx = 0$

(5) $\dfrac{\cos y}{\sin y} dy + \dfrac{\cos x}{\sin x} dx = 0$

(6) Not separable

(7) $\dfrac{y\, dy}{1+y^2} = \dfrac{dx}{x(1+x^2)}$

FRAME 20

Technique

To complete the solution of this type of equation we simply integrate the separated result. Thus, returning to FRAME 18, we have

$$\int \frac{dy}{g(y)} = \int f(x)\,dx + A$$

where A is an arbitrary constant of integration.

You may consider that 2 constants of integration are required, one for each integral, yielding

$$\int \frac{dy}{g(y)} + A_1 = \int f(x)\,dx + A_2, \quad \text{say.}$$

However, you will recall from FRAME 8 that only <u>one</u> constant is needed, hence A_1 may be combined with A_2 giving a new constant $A = A_2 - A_1$.

FRAME 21

Let us consider the separated equation in FRAME 18A.

$$\frac{dy}{y} = (x^2 + 1)dx$$

Integrating we have

$$\int \frac{dy}{y} = \int (x^2 + 1)dx + A$$

i.e.
$$\log_e y = \tfrac{1}{3}x^3 + x + A$$

which is the general solution to the d.e.

FRAME 22

In the above general solution the form of the arbitrary constant <u>could be</u> chosen to be $\log_e B$ for convenience.

Thus,
$$\log_e y = \tfrac{1}{3}x^3 + x + \log_e B$$

or
$$\log_e y - \log_e B = x + \tfrac{1}{3}x^3$$

or
$$\log_e \tfrac{y}{B} = x + \tfrac{1}{3}x^3.$$

Taking anti-logs we have

$$\frac{y}{B} = e^{x+x^3/3}$$

yielding
$$y = Be^{x+x^3/3}$$

as an alternative and possibly more convenient form of general solution.

FRAME 23

The general solutions obtained in FRAMES 21 and 22, though different in form, are entirely equivalent.

You will find in solving d.e.'s that the general solution to a d.e. may be obtained in different forms by a suitable choice of the arbitrary constants.

FIRST ORDER DIFFERENTIAL EQUATIONS

FRAME 23 continued

With these remarks in mind, go on to complete the solutions of the "separable variable" equations in FRAME 19.

****

23A

(1) $\tan y = \tan x + A$

(2) $y = \dfrac{x + A}{1 - Ax}$ (Hint: choose arbitrary constant to be $\tan^{-1} A$.)

(4) $y = Ax \exp\left(\dfrac{1}{x} + \dfrac{1}{y}\right)$

(5) $\sin x \sin y = A$

(7) $(1 + y^2) = \dfrac{Ax^2}{1 + x^2}$

(Integration after partial fractions yields

$$\tfrac{1}{2} \log_e(1 + y^2) = \log_e x - \tfrac{1}{2} \log_e(1 + x^2) + \log_e B.$$

Result follows with $A = B^2$.)

FRAME 24

Find the particular solutions in the above cases with the following boundary conditions respectively:

(1) $y = \dfrac{\pi}{4}$ when $x = 0$

(2) $y = 1$ when $x = 0$

(4) $y = 1$ when $x = 1$

(5) $y = \dfrac{\pi}{2}$ when $x = \dfrac{\pi}{3}$

(7) $y = 0$ when $x = 1$

****

  24A

(1) $\tan y = 1 + \tan x$

(2) $y = \dfrac{1 + x}{1 - x}$

(4) $y = x \exp\left(\dfrac{1}{x} + \dfrac{1}{y} - 2\right)$

(5) $2 \sin x \sin y = \sqrt{3}$

(7) $y^2 = \dfrac{x^2 - 1}{x^2 + 1}$

                                                          FRAME 25

For extra practice, perhaps you would like to solve the following equations:

(1) $\sqrt{1 + x^2}\,\dfrac{dy}{dx} = xe^y$, given that $y = 0$ when $x = \dfrac{3}{4}$

(2) $x\,\dfrac{dy}{dx} - y = xy$, given that $y = 1$ when $x = 2$

(3) $\operatorname{sech} x\,\dfrac{dy}{dx} = e^{y-x}$

(4) $x^3\,\dfrac{dy}{dx} + 3y^2 = xy^2$, subject to the condition that $y = 1$ when $x = 1$

(5) $r\,\dfrac{d\theta}{dr} = \dfrac{c^2 - r^2}{c^2 + r^2}\tan\theta$, given that $r = c$ when $\theta = \dfrac{\pi}{2}$.

                ****
                                                          25A

(1) $\sqrt{1 + x^2} + e^{-y} = \dfrac{9}{4}$

(2) $y = \dfrac{1}{2}xe^{x-2}$

(3) $4e^{-y} = e^{-2x} - 2x + A$

(4) $2x^2 - 2xy + 3y - 3x^2y = 0$

(5) $\sin\theta = \dfrac{2rc}{r^2 + c^2}$

FIRST ORDER DIFFERENTIAL EQUATIONS 1:13

FRAME 26

Special cases of Type I d.e.'s

Specially simple cases of the equation $\frac{dy}{dx} = f(x)g(y)$ arise when $f(x)$ or $g(y)$ are constants. Solve the following examples of such cases.

(1) $\quad \frac{dy}{dx} = -ky$, given that $y = y_0$ when $x = 0$

(2) $\quad \frac{dy}{dx} = 2xe^{-x}$

(3) $\quad \frac{dy}{dx} = y^2 + 1$

****

26A

(1) $\quad y = y_0 e^{-kx}$

(2) $\quad y = -2(x + 1)e^{-x} + A$

(3) $\quad y = \tan(x + A)$

FRAME 27

Reduction to Type I by substitution

It is sometimes possible by means of a simple substitution to reduce an equation in which the variables cannot be separated to one where they can.

Consider $\frac{dy}{dx} = \cos(x + y)$ in which the variables are not separable.

Make the substitution $x + y = z$, say.

$$\text{Then } 1 + \frac{dy}{dx} = \frac{dz}{dx}.$$

Substitute for $\frac{dy}{dx}$ to obtain an equation in z and x. Is this an equation of Type I? If so, separate the variables, integrate and finally re-substitute for z to obtain the general solution.

****

27A

$\frac{dz}{dx} - 1 = \cos z$ in which the variables are separable.

$\therefore \int \frac{dz}{1 + \cos z} = \int dx + A$

Integrating, $\tan \frac{z}{2} = x + A$ {NOTE: $1 + \cos z = 2\cos^2(z/2)$}

whence $y = 2\tan^{-1}(x + A) - x$.

FRAME 28

Solve the following equations:

(1) $\quad \frac{dy}{dx} = \tan^2(x + y)$

(2) $\quad \frac{dy}{dx} = (x + 4y)^2$

****

28A

(1) $\quad \sin 2(x + y) = 2x - 2y + A$

(2) $\quad y = \frac{1}{8}\{\tan(2x + A) - 2x\}$

FRAME 29

Applications

We shall now consider some practical examples arising from science and engineering.

FRAME 30

Example 1

In a certain termolecular reaction the decrease x in concentration of a substance A is given by $\frac{dx}{dt} = k(a - x)^3$ where k is the reaction rate constant and a is the initial concentration of A. Find the concentration at time t.

FIRST ORDER DIFFERENTIAL EQUATIONS

FRAME 30 continued

Separating the variables, you will obtain $\dfrac{dx}{(a-x)^3} = k\,dt$.

Integrating,
$$\dfrac{1}{2(a-x)^2} = kt + B. \qquad (B \text{ constant})$$

Now find the arbitrary constant B by applying the initial condition that the concentration decrease x is zero at $t = 0$.

From the particular solution thus obtained, find the concentration $(a-x)$.

****

30A

Using $x = 0$ at $t = 0$ gives

$$\dfrac{1}{2(a-x)^2} = kt + \dfrac{1}{2a^2}$$

whence,   concentration $= (a-x) = \dfrac{a}{\sqrt{1 + 2a^2 kt}}$.

FRAME 31

Example 2

A circuit consists of a resistance $R\,\Omega$ and an inductance L H connected in series to a battery of constant voltage E_o. Find the current, i A, at time t s after closing the circuit.

Equating the total potential drop across each of the components to E_o we have

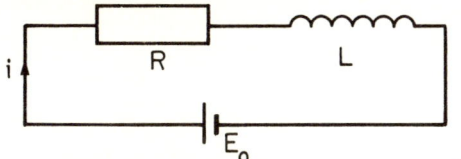

the d.e. $L\dfrac{di}{dt} + Ri = E_o$

subject to the initial condition $i = 0$ when $t = 0$.

Separating the variables $\quad dt = \dfrac{L\,di}{E_o - Ri}$

which yields $\quad t = -\dfrac{L}{R}\log_e(E_o - Ri) + \dfrac{L}{R}\log_e A.$

You will note that for algebraic convenience the arbitrary constant has been chosen in the above form.

Apply the initial condition and find i from the result.

****

$$t = -\frac{L}{R} \log_e \frac{E_o - Ri}{E_o}$$

giving $\quad i = \frac{E_o}{R} (1 - e^{-Rt/L})$

FRAME 32

Alternatively, the determination of the arbitrary constant may be avoided by incorporating the boundary condition in the form of definite integrals. Thus in FRAME 31

$$\int_o^t dt = \int_o^i \frac{L\,di}{E_o - Ri}$$

$\therefore \quad t = \left[-\frac{L}{R} \log_e (E_o - Ri) \right]_o^i$

i.e. $\quad t = -\frac{L}{R} \log_e \frac{E_o - Ri}{E_o} \quad$ as above.

This technique is frequently employed in practical examples.

FRAME 33

Example 3

Find the velocity with which a particle must be projected from the surface of the earth to escape from the earth's gravitational field. (Neglect the resistance of the atmosphere.)

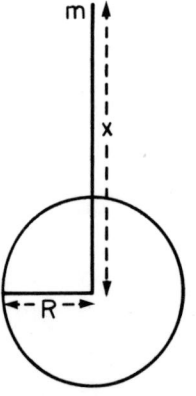

The gravitational attraction of the earth varies as $1/x^2$ and is equal to mg when $x = R$. Hence, in general it is

$$\frac{mgR^2}{x^2}.$$

$\therefore$ Equation of motion is

$$mv\frac{dv}{dx} = -\frac{mgR^2}{x^2}$$

FIRST ORDER DIFFERENTIAL EQUATIONS

FRAME 33 continued

Separate the variables to obtain

$$v \, dv = -gR^2 \frac{dx}{x^2}.$$

Integrate and apply the boundary condition that when $x = R$, $v = V$ (the velocity of projection).

****

33A

$$\tfrac{1}{2}v^2 = \tfrac{1}{2}V^2 + \frac{gR^2}{x} - gR$$

FRAME 34

For the particle to escape, v must still be positive even when x is large. Hence,

$$\tfrac{1}{2}V^2 - gR \geq 0$$

and the escape velocity is $V = \sqrt{2gR}$.

Taking $g = 9 \cdot 81 \text{ m/s}^2$ and $R = 6378$ km

escape velocity $= 11 \cdot 2$ km/s.

FRAME 35

If you feel that you require some extra practice, some further examples on various applications are given in this frame and in FRAMES 36 - 38.

Example 4

The rate at which a radio-active substance decays is proportional to the number of atoms N present at time t. If the constant of proportionality is λ (the decay constant) and initially there are $\dot{N_o}$ atoms present, express N as a function of t.

Find also the time taken for half of the initial amount to decay.

****

35A

$$\frac{dN}{dt} = -\lambda N$$

$$N = N_0 e^{-\lambda t}$$

$$N = N_0/2 \quad \text{when} \quad t = (\log_e 2)/\lambda.$$

FRAME 36

Example 5

A particle falls from rest at a great height H m above the earth. Neglecting air resistance, prove that it reaches the surface of the earth with velocity $\left(\frac{2gRH}{R+H}\right)^{\frac{1}{2}}$ m/s where R m is the radius of the earth.

(The d.e. is as in FRAME 33.)

****

36A

Equation is $\quad v\dfrac{dv}{dx} = -\dfrac{gR^2}{x^2} \quad$ subject to $\quad v = 0 \quad$ when $\quad x = R + H.$

Solution is $\quad \dfrac{1}{2}v^2 = gR^2\left(\dfrac{1}{x} - \dfrac{1}{R+H}\right)$

v is required when $x = R.$

FRAME 37

Example 6

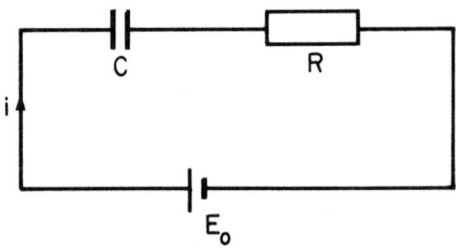

A condenser of capacitance C F is charged through a resistance R Ω by a battery of constant voltage E_0. The charge Q on the condenser at time t is given by the equation

$$\frac{Q}{C} + Ri = E_0$$

where i A is the current flowing in the circuit.

Noting that $i = \dfrac{dQ}{dt}$ and that initially $Q = 0$, find the charge Q at time t and the value of Q when the condenser is fully charged.

****

FIRST ORDER DIFFERENTIAL EQUATIONS

37A

Equation is $R \frac{dQ}{dt} = E_o - \frac{Q}{C}$.

Solution is $Q = E_o C(1 - e^{-t/RC})$.

Ultimate value of Q is $E_o C$ (i.e. as $t \to \infty$).

FRAME 38

Example 7

A sphere of mass m and radius a falls from rest under gravity in a fluid whose resistance to motion is given by Stokes' Law to be $6\pi\eta a v$, η being the fluid viscosity and v the speed of the particle. Show that the particle ultimately attains a speed of $mg/6\pi\eta a$ (its terminal velocity), and find the time taken to acquire one half of this speed.

****

38A

Equation of motion is $m \frac{dv}{dt} = mg - mkv$

where $k = 6\pi\eta a/m$.

Solution, subject to $v = 0$ at $t = 0$ is

$$v = \frac{g}{k}(1 - e^{-kt}).$$

Terminal velocity is $V = g/k$ (i.e. as $t \to \infty$) and $v = V/2$ when $t = \frac{V}{g} \log_e 2$.

FRAME 39

Type II - Homogeneous

Recognition

M(x,y) is said to be a homogeneous function of degree n if the sum of the powers of x and y in <u>each</u> term of M is n.

e.g.   (i) $x^2 y - 3xy^2 + 2y^3$ is homogeneous of degree 3

   (ii) $x^4 - 7x^2 y^2$ is homogeneous of degree 4.

FRAME 39 continued

If a first order d.e. is written in the form

$$\frac{dy}{dx} = \frac{M(x,y)}{N(x,y)}$$

where M and N are homogeneous functions of the same degree, then the equation is said to be HOMOGENEOUS.

e.g. $\quad \dfrac{dy}{dx} = \dfrac{xy}{x^2 + y^2} \quad$ is homogeneous.

FRAME 40

Which of the following d.e.'s are homogeneous?

(1) $\quad (x^2 + y^2)\dfrac{dy}{dx} = xy$

(2) $\quad x(x^2 + 3y^2)\dfrac{dy}{dx} = y(3x^2 + y^2)$

(3) $\quad (x^2 + y)\dfrac{dy}{dx} = xy$

(4) $\quad \dfrac{dy}{dx} = \dfrac{x^2 + 2y^2}{xy}$

(5) $\quad \dfrac{dy}{dx} = \dfrac{y(3x^2 + y^2)}{x(x + 3y)}$

(6) $\quad \dfrac{dy}{dx} = \dfrac{y(x + 2y)}{x(2x + y)}$

****

40A

(1), (2), (4) and (6) are homogeneous.

(3) and (5) are not homogeneous.

In (3) note that $x^2 + y$ is not homogeneous.

In (5) numerator and denominator are both homogeneous functions but are not of the same degree.

FIRST ORDER DIFFERENTIAL EQUATIONS

FRAME 41

Technique

If, in the equation $\dfrac{dy}{dx} = \dfrac{M(x,y)}{N(x,y)}$, both M and N are homogeneous of degree n, we may divide them by x^n and express the R.H.S. as a function of the single variable v, where $v = \dfrac{y}{x}$.

For example, in equations (1) and (2) of FRAME 40 we have:

(1) $\quad \dfrac{dy}{dx} = \dfrac{xy}{x^2 + y^2} = \dfrac{\frac{y}{x}}{1 + \left(\frac{y}{x}\right)^2} = \dfrac{v}{1 + v^2}$,

   M and N both being divided by x^2,

and

(2) $\quad \dfrac{dy}{dx} = \dfrac{y(3x^2 + y^2)}{x(x^2 + 3y^2)} = \dfrac{v(3 + v^2)}{1 + 3v^2}$,

   division this time being by x^3.

Confirm that in equations (3) and (5) this technique will not be successful in that the R.H.S.'s cannot be expressed in terms of v alone.

FRAME 42

The previous frame suggests the use of the substitution $y/x = v$ or $y = vx$ as the standard method of solving homogeneous d.e.'s.

Hence $\dfrac{dy}{dx} = v + x\dfrac{dv}{dx}$, and substituting this in the d.e. you will see that the result is a new d.e. in which the variables v and x can be separated.

Having done this, you can then integrate to obtain a relation between v and x and, finally, replace v by y/x.

FRAME 43

For example, returning to equation (1) in FRAME 40 and using the technique suggested in FRAMES 41 and 42 the equation $(x^2 + y^2)\dfrac{dy}{dx} = xy$ becomes

$$v + x\dfrac{dv}{dx} = \dfrac{v}{1 + v^2}$$

i.e. $\quad x\dfrac{dv}{dx} = \dfrac{v}{1+v^2} - v = -\dfrac{v^3}{1+v^2}$

in which v and x are separable.

Separate the variables, integrate and substitute for v to obtain the general solution.

****
                                                                43A

$$\dfrac{dx}{x} = \left(-\dfrac{1}{v^3} - \dfrac{1}{v}\right)dv$$

$$\therefore \log_e x = \dfrac{1}{2v^2} - \log_e v + \log_e A.$$

$$\text{Hence } \log_e \dfrac{y}{A} = \dfrac{x^2}{2y^2}.$$

FRAME 44

In a similar manner, equation (2) of FRAME 40 may be written as

$$\dfrac{dy}{dx} = \dfrac{y(3x^2 + y^2)}{x(x^2 + 3y^2)}$$

and becomes

$$v + x\dfrac{dv}{dx} = \dfrac{v(3 + v^2)}{1 + 3v^2}.$$

Complete the solution as above.

****
                                                                44A

$$\dfrac{dx}{x} = \dfrac{1 + 3v^2}{2v(1 - v^2)}\,dv$$

$$= \dfrac{1}{2}\left(\dfrac{1}{v} + \dfrac{2}{1-v} - \dfrac{2}{1+v}\right)dv \quad \textit{on splitting into partial fractions.}$$

Integrating and combining you will obtain

$$\log_e x^2 = \log_e \dfrac{Av}{(1-v^2)^2}$$

whence $\qquad (x^2 - y^2)^2 = Axy.$

FIRST ORDER DIFFERENTIAL EQUATIONS

FRAME 45

Now try some or all of the following equations:

(1) $\dfrac{dy}{dx} = \dfrac{x^2 + 2y^2}{xy}$ given that $y = 0$ when $x = 1$

(2) $\dfrac{dy}{dx} = \dfrac{y(x + 2y)}{x(2x + y)}$

(3) $\dfrac{dy}{dx} = \dfrac{y}{x} + \tan \dfrac{y}{x}$

(4) $(x^4 + y^4) = 2x^3 y \dfrac{dy}{dx}$

(5) $(x^2 + 2xy)\dfrac{dy}{dx} + 2xy + y^2 + 3x^2 = 0$

given that $y = 2$ when $x = 1$.

****

45A

(1) $y^2 = x^4 - x^2$

(2) $x^2 y^2 = A(y - x)^3$

(3) $x = A \sin \dfrac{y}{x}$

(4) $x^2 = (x^2 - y^2) \log_e Ax$

(5) $x(x^2 + xy + y^2) = 7.$

FRAME 46

Application

A particle of mass m moves along the x-axis and is attracted towards the origin O by a force $mn^2 x$ proportional to its displacement from O.

It is also subject to a frictional resistance of magnitude mkv which varies as the velocity v.

Obtain a general expression relating velocity and displacement assuming that $k^2 < 4n^2$.

FIRST ORDER DIFFERENTIAL EQUATIONS

FRAME 46 continued

The equation of motion is
$$mv \frac{dv}{dx} = -mkv - mn^2 x,$$
which is homogeneous in v and x.

Substitute $v/x = u$, say, or $v = xu$ which gives
$$\frac{dv}{dx} = u + x\frac{du}{dx}$$
and verify that the equation becomes
$$u + x\frac{du}{dx} = -\frac{ku + n^2}{u}$$
i.e.
$$\frac{dx}{x} = -\frac{u\,du}{u^2 + ku + n^2}.$$

Complete the integration and replace u by v/x.

****

46A

$$-\frac{dx}{x} = \left\{ \frac{(2u + k)/2}{u^2 + ku + n^2} - \frac{k/2}{\left(u + \frac{k}{2}\right)^2 + \alpha^2} \right\} du$$

where $\alpha^2 = n^2 - \frac{k^2}{4}$

$\therefore \; -\log_e x = \frac{1}{2}\log_e(u^2 + ku + n^2) - \frac{k}{2\alpha}\tan^{-1}\left(\frac{u}{\alpha} + \frac{k}{2\alpha}\right) + constant.$

This yields $\quad \log_e(v^2 + kvx + n^2 x^2) - \frac{k}{\alpha}\tan^{-1}\left(\frac{v}{\alpha x} + \frac{k}{2\alpha}\right) = A.$

FRAME 47

Equations reducible to homogeneous

The equation $\frac{dy}{dx} = \frac{ax + by + c}{a'x + b'y + c'}$ is not homogeneous, but can be reduced to this type by the substitutions
$$Y = ax + by + c$$
$$X = a'x + b'y + c'.$$

Verify that the equation reduces to
$$\frac{dY}{dX} = \frac{aX + bY}{a'X + b'Y} \quad \text{which is homogeneous.}$$

****

FIRST ORDER DIFFERENTIAL EQUATIONS

47A

$$dY = a\,dx + b\,dy$$
$$dX = a'\,dx + b'\,dy.$$

These can be solved simultaneously for dx and dy. $\dfrac{dY}{dX}$ can then be formed and the result follows on noting that $\dfrac{dY}{dX} = \dfrac{Y}{X}$.

FRAME 48

Example

$$\frac{dy}{dx} = \frac{x - y + 2}{x + y - 2}$$

This reduces to the homogeneous equation $\dfrac{dY}{dX} = \dfrac{X - Y}{X + Y}$ where $Y = x - y + 2$ and $X = x + y - 2$.

Complete the solution by using the substitution
$$Y/X = V, \quad \text{say.}$$

****

48A

$$-\frac{dX}{X} = \frac{V + 1}{V^2 + 2V - 1}\,dV$$

$\therefore\ \log_e A - \log_e X = \tfrac{1}{2}\log_e(V^2 + 2V - 1)$

from which

$$Y^2 + 2XY - X^2 = B. \qquad (B = A^2)$$

Re-substituting,

$$(x - y + 2)^2 + 2(x - y + 2)(x + y - 2) - (x + y - 2)^2 = B.$$

This could be further simplified, if desired.

FRAME 49

Now try the following examples:

(1) $\dfrac{dy}{dx} = \dfrac{2x - 5y + 3}{2x + 4y - 6}$

(2) $\dfrac{dy}{dx} = \dfrac{2x - 2y + 3}{x - y + 1}$.

Note that in the latter case it is not necessary (nor indeed possible!) to use X for the denominator <u>and</u> Y for the numerator. The reason is that the coefficients of x and y in the numerator and denominator respectively are in proportion and consequently Y can be expressed linearly in terms of X. So, choose $X = x - y + 1$ and hence $2x - 2y + 3 = 2X + 1$.

$\dfrac{dy}{dx}$ is then replaced by $1 - \dfrac{dX}{dx}$ and the result is an equation in X and x in which the variables can be separated.

****

49A

(1) $(x - 4y + 3)(2x + y - 3)^2 = A$

(2)   *The separated equation is* $\dfrac{XdX}{X + 1} = -dx$

and yields $(2x - y + 1) = \log_e A(x - y + 2)$.

FRAME 50

Type III - Linear

Recognition

If a d.e. can be written in the form $\dfrac{dy}{dx} + Py = Q$ where P and Q are functions of x only, the equation is said to be LINEAR of the first order, since $\dfrac{dy}{dx}$ and y occur linearly.

e.g. $\dfrac{dy}{dx} + 2y \cot x = \cos x$ is linear with $P = 2 \cot x$, $Q = \cos x$,

and $x\dfrac{dy}{dx} - 3y = x - 1$

may be written $\dfrac{dy}{dx} - \dfrac{3}{x}y = 1 - \dfrac{1}{x}$

which is again linear with $P = -\dfrac{3}{x}$ and $Q = 1 - \dfrac{1}{x}$.

FIRST ORDER DIFFERENTIAL EQUATIONS

FRAME 51

Which of the following equations are linear?
Do you recognise the types of the other equations?

(1) $\quad x^2 \dfrac{dy}{dx} - (x + 1)y = \sin x$

(2) $\quad x^2 \dfrac{dy}{dx} - y^2 = xy$

(3) $\quad \dfrac{dy}{dx} + \dfrac{x}{1 + x^2} y = \dfrac{1}{2x(1 + x^2)}$

(4) $\quad \sec^2 x \tan y \, dx + \sec^2 y \tan x \, dy = 0$

(5) $\quad \dfrac{dy}{dx} = \dfrac{x + y}{x}$

(6) $\quad (x^2 + 1)\dfrac{dy}{dx} = y + 1$

****

51A

(1) Linear with $P = -\dfrac{x + 1}{x^2}$ and $Q = \dfrac{\sin x}{x^2}$

(2) Homogeneous

(3) Linear with $P = \dfrac{x}{1 + x^2}$ and $Q = \dfrac{1}{2x(1 + x^2)}$

(4) Variables separable

(5) Linear with $P = -\dfrac{1}{x}$ and $Q = 1$ and also <u>homogeneous</u>

(6) Linear with $P = -\dfrac{1}{x^2 + 1}$ and $Q = \dfrac{1}{x^2 + 1}$ and <u>also</u> variables separable

---

FRAME 52

Technique

In the standard linear equation
$$\dfrac{dy}{dx} + Py = Q,$$
the presence of the terms $\dfrac{dy}{dx}$ and y suggests the differentiation of a product involving y.

To produce this product we multiply the equation throughout by a function of x, u say, to be determined later.

FIRST ORDER DIFFERENTIAL EQUATIONS

FRAME 52 continued

This gives
$$u \frac{dy}{dx} + uPy = uQ$$
which may be written as
$$\frac{d}{dx}(uy) + (uPy - y\frac{du}{dx}) = uQ.$$

This equation could be solved by direct integration if $uPy - y\frac{du}{dx}$ vanished since the equation would then reduce to
$$\frac{d}{dx}(uy) = uQ. \qquad (52.1)$$

Hence we impose the condition
$$uPy - y\frac{du}{dx} = 0,$$
i.e.
$$\frac{du}{dx} = uP,$$
since $y = 0$ is not acceptable as a solution.

This is a variables separable d.e. in u and x.

Solve it to obtain u.

****

52A

$$\frac{du}{u} = P dx$$

$$\log_e u = \int P dx$$

$$u = e^{\int P dx}$$

Note that no arbitrary constant need be included here since the one constant required in the solution of the original d.e. will arise on performing the integration of equation (52.1). This function u is referred to as the INTEGRATING FACTOR (I.F.).

FIRST ORDER DIFFERENTIAL EQUATIONS

FRAME 53

In solving linear equations you may find it useful to refer to the following summary of the steps involved.

(i) Write the equation in standard form $\frac{dy}{dx} + Py = Q$.

(ii) Evaluate $\int P dx$ and obtain the integrating factor $u = e^{\int P dx}$. (No arbitrary constant is introduced at this stage.)

(iii) Multiply the equation in (i) by u and check that the result is $\frac{d}{dx}(uy) = uQ$.

(iv) Integrate to complete the solution as $uy = \int uQ dx + A$.

FRAME 54

Example

Solve $\quad x\frac{dy}{dx} + 2y = e^x$.

Writing this in standard form

$$\frac{dy}{dx} + \frac{2}{x}y = \frac{e^x}{x}, \qquad (54.1)$$

you will note that this is linear with $P = \frac{2}{x}$.

$\therefore \quad \int P dx = 2 \log_e x = \log_e x^2$.

The I.F. $u = e^{\int P dx} = e^{\log_e x^2} = x^2$.

Now, multiply (54.1) by u and check that the L.H.S. becomes $\frac{d}{dx}(x^2 y)$.

Finally, complete the solution by integrating.

****

54A

$$x^2 \frac{dy}{dx} + 2xy = xe^x$$

$$\therefore \frac{d}{dx}(x^2 y) = xe^x. \qquad (Check!)$$

Integrating, $\quad x^2 y = \int xe^x dx + A$

yielding $\quad y = \frac{e^x}{x^2}(x - 1) + \frac{A}{x^2}$

FRAME 55

In evaluating the I.F. $u = e^{\int P dx}$, the integral $\int P dx$ frequently involves logarithmic expressions. You will find it useful to note that
$$e^{\log_e f(x)} \equiv \exp \log_e f(x) = f(x).$$

For example, if $P = -\frac{2}{3} \cot 2x$, then
$$\int P dx = -\frac{1}{3} \log_e \sin 2x = \log_e (\operatorname{cosec} 2x)^{1/3},$$
and $u = e^{\int P dx} = (\operatorname{cosec} 2x)^{1/3}.$

Again, if $P = \frac{x+2}{x+1}$, then
$$\int P dx = \int (1 + \frac{1}{x+1}) dx = x + \log_e (x+1)$$
$$\therefore u = e^{x + \log_e (x+1)} = e^x (x+1).$$

Obtain the integrating factors corresponding to the following expressions for P:

(1) $-\frac{3}{x}$ (2) $\frac{x^3}{1+x^4}$

(3) $\tan x - \frac{1}{x}$ (4) $\frac{x}{x-1}$

(5) $\frac{1}{x(x^2 - 1)}$

****

55A

(1) $\frac{1}{x^3}$ (2) $(1 + x^4)^{1/4}$

(3) $\frac{\sec x}{x}$ (4) $(x-1)e^x$

(5) $\frac{\sqrt{x^2 - 1}}{x}$

$$\left[\text{Note that: } \frac{1}{x(x^2-1)} \equiv \frac{1}{2(x+1)} + \frac{1}{2(x-1)} - \frac{1}{x} \right]$$

FIRST ORDER DIFFERENTIAL EQUATIONS

FRAME 56

Example

Let us now return to a more difficult linear d.e.

Solve $x(x^2 - 1)\dfrac{dy}{dx} + y = x^3$.

Written in standard form this is

$$\dfrac{dy}{dx} + \dfrac{1}{x(x^2 - 1)} y = \dfrac{x^2}{x^2 - 1} \qquad (56.1)$$

which shows that $P = \dfrac{1}{x(x^2 - 1)}$.

You have already integrated this expression in practice example (5) of FRAME 55 and obtained the integrating factor $u = \dfrac{\sqrt{x^2 - 1}}{x}$.

Multiply (56.1) by u and check that the L.H.S. becomes $\dfrac{d}{dx}\left(\dfrac{\sqrt{x^2 - 1}}{x} y\right)$.

Integrate to complete the solution.

****

56A

$$\dfrac{\sqrt{x^2 - 1}}{x} \dfrac{dy}{dx} + \dfrac{1}{x^2 \sqrt{x^2 - 1}} \cdot y = \dfrac{x}{\sqrt{x^2 - 1}}$$

$\therefore \dfrac{d}{dx}\left(\dfrac{\sqrt{x^2 - 1}}{x} y\right) = \dfrac{x}{\sqrt{x^2 - 1}} \qquad$ (Check!).

Integrating,

$$\dfrac{\sqrt{x^2 - 1}}{x} y = \sqrt{x^2 - 1} + A$$

i.e. $\qquad y = x + \dfrac{Ax}{\sqrt{x^2 - 1}}$.

FRAME 57

Solve the following practice examples:

(1) $(x^2 + 1)\dfrac{dy}{dx} + 2xy = 4x^2$, given that $y = 4$ when $x = 3$

(2) $x\dfrac{dy}{dx} - y = 2x^2 \operatorname{cosec} 2x$, given that $y = 1$ when $x = \dfrac{\pi}{4}$

(3) $(1 + x^2)\dfrac{dy}{dx} - xy = x(1 + x^2)$

FRAME 57 continued

(4) $\quad x \cos x \dfrac{dy}{dx} + (x \sin x - \cos x)y - x^2 = 0$,
given that $y = 0$ when $x = \dfrac{\pi}{4}$

(5) $\quad x(1 - x^2)\dfrac{dy}{dx} + (2x^2 - 1)y = x^3$

****

57A

(1) $\quad$ I.F. $= (x^2 + 1); \quad y(x^2 + 1) = \dfrac{4}{3}x^3 + 4$

(2) $\quad$ I.F. $= \dfrac{1}{x}; \quad y = x(\dfrac{4}{\pi} + \log_e \tan x)$

(3) $\quad$ I.F. $= \dfrac{1}{\sqrt{1 + x^2}}; \quad y = (1 + x^2) + A\sqrt{1 + x^2}$

(4) $\quad$ I.F. $= \dfrac{\sec x}{x}; \quad y = x(\sin x - \cos x)$

(5) $\quad$ I.F. $= \dfrac{1}{x\sqrt{1 - x^2}}; \quad y = x + Ax\sqrt{1 - x^2}$

FRAME 58

Applications

Example

Return to the R,L series circuit of FRAME 31 in which the d.e. for the current i at time t is
$$L \dfrac{di}{dt} + Ri = E_o,$$
subject to the initial condition $i = 0$ when $t = 0$.

You will recall that this equation was solved by separating the variables. You will note that this equation is also <u>linear</u> and it will now be solved using the technique for such equations. In addition the equation will also be solved in the case when the constant voltage E_o is replaced by the alternating voltage $E_o \sin \omega t$. The equations are then:

$$(i) \quad L \dfrac{di}{dt} + Ri = E_o,$$
$$(ii) \quad L \dfrac{di}{dt} + Ri = E_o \sin \omega t.$$

Incidentally, (ii) must be solved using the linear method since the variables cannot be separated.

FIRST ORDER DIFFERENTIAL EQUATIONS

FRAME 59

Writing equation (i) in standard form,
$$\frac{di}{dt} + \frac{R}{L}i = \frac{E_o}{L}$$

and $\therefore$ $P = \frac{R}{L}$ and $\int P\,dt = \frac{Rt}{L}$.

Hence the I.F. is $e^{Rt/L}$.

Multiply by this factor and complete the solution in the usual way, finding the arbitrary constant by applying the initial condition.

****

59A

$$\frac{d}{dt}(ie^{Rt/L}) = \frac{E_o}{L}e^{Rt/L}$$

$$\therefore \quad ie^{Rt/L} = \frac{E_o}{R}e^{Rt/L} + A$$

giving $\quad i = \frac{E_o}{R}(1 - e^{-Rt/L})$.

FRAME 60

In standard form (ii) becomes
$$\frac{di}{dt} + \frac{R}{L}i = \frac{E_o}{L}\sin \omega t$$

The I.F. is $e^{Rt/L}$ as for (i). Complete the solution in the usual way.

****

60A

$$\frac{d}{dt}(ie^{Rt/L}) = \frac{E_o}{L}e^{Rt/L}\sin \omega t.$$

Integration yields
$$ie^{Rt/L} = \frac{E_o}{L}\frac{e^{Rt/L}\left[\frac{R}{L}\sin \omega t - \omega \cos \omega t\right]}{\frac{R^2}{L^2} + \omega^2} + A.$$

Using the initial condition, it follows that
$$i = \frac{E_o}{R^2 + \omega^2 L^2}(R \sin \omega t - \omega L \cos \omega t + \omega L e^{-Rt/L}).$$

FRAME 61

Further examples on practical applications for you to solve are given here and in FRAMES 62 and 63.

Example 1

In an R,C series circuit in which there is an applied voltage E which may vary with time t, the d.e. for the charge Q on the condenser at time t is

$$R \frac{dQ}{dt} + \frac{Q}{C} = E$$

with $Q = 0$ initially, as in FRAME 37. Find Q in the cases when:

    (i) $E = E_o$,
    (ii) $E = E_o \sin \omega t$.

(Note that case (i) was treated by separation of variables in FRAME 37.)

****

61A

(i) $Q = E_o C (1 - e^{-t/RC})$

(ii) $Q = \dfrac{E_o C}{1 + \omega^2 R^2 C^2}(\sin \omega t - \omega RC \cos \omega t + \omega RC e^{-t/RC})$

FRAME 62

Example 2

A coil of n turns of area A whose inductance is L and resistance R is rotated with angular velocity ω in a uniform magnetic field of strength H about a diameter at right angles to the field. The current i induced in the coil is given by the d.e.

$$L \frac{di}{dt} + Ri = n\omega HA \cos \omega t.$$

Find the current at time t, assuming that initially it is zero.

****

62A

$$i = \frac{n\omega HA}{R^2 + \omega^2 L^2}(R \cos \omega t + \omega L \sin \omega t - R e^{-Rt/L}).$$

FIRST ORDER DIFFERENTIAL EQUATIONS

FRAME 63

Example 3

A particle of mass m moves along the x-axis and is attracted towards the origin O by a force mn^2x proportional to its displacement from O. It is also subject to a frictional resistance of magnitude mkv^2 which varies as the square of the velocity v (cf. FRAME 46). If the particle starts from rest at distance a from O show that it reaches O with velocity

$$\left\{\frac{n^2}{2k^2}(1 - e^{2ka} + 2kae^{2ka})\right\}^{\frac{1}{2}}$$

(Hint: The d.e. of motion which you will obtain can be reduced to the linear type by means of the substitution $y = v^2$.)

****

63A

Equation of motion is

$$mv\frac{dv}{dx} = -mkv^2 - mn^2x$$

which reduces to the linear equation

$$\frac{dy}{dx} + 2ky = -2n^2x. \qquad (y = v^2)$$

General solution is

$$v^2 = \frac{n^2}{2k^2}(1 - 2kx) + Ae^{-2kx}.$$

A is found by noting that $v = 0$ when $x = a$ and then v is obtained when $x = 0$.

FRAME 64

Type IV - Exact

Recognition

The first order d.e.
$$P(x,y)dx + Q(x,y)dy = 0 \qquad (64.1)$$

is said to be **exact** if a function $u(x,y)$ exists such that
$$du = Pdx + Qdy. \qquad (64.2)$$

FRAME 64 continued

If such a function exists the equation becomes $du = 0$ and is immediately integrable to give $u = A$ where A is a constant.

For instance, consider the d.e.
$$3x^2y^2 dx + 2x^3y \, dy = 0.$$

By inspection, this can be written as
$$du = d(x^3y^2) = 0$$

with solution $\quad x^3y^2 = A.$

FRAME 65

Since $\quad du = \dfrac{\partial u}{\partial x} dx + \dfrac{\partial u}{\partial y} dy,$

comparison with (64.2) gives
$$\frac{\partial u}{\partial x} = P \quad \text{and} \quad \frac{\partial u}{\partial y} = Q. \qquad (65.1)$$

Then, noting that
$$\frac{\partial^2 u}{\partial y \partial x} = \frac{\partial^2 u}{\partial x \partial y}, \quad \text{we obtain}$$

$$\frac{\partial P}{\partial y} = \frac{\partial Q}{\partial x} \qquad (65.2)$$

This is the condition for $Pdx + Qdy$ to be an exact differential.

Example

In the equation
$$(1 + 4xy + 2y^2)dx + (1 + 4xy + 2x^2)dy = 0$$
verify that
$$\frac{\partial P}{\partial y} = \frac{\partial Q}{\partial x}$$
and hence that the equation is exact.

****

65A

$$\frac{\partial P}{\partial y} = 4x + 4y = \frac{\partial Q}{\partial x}$$

FIRST ORDER DIFFERENTIAL EQUATIONS

FRAME 66

Which of the following equations are exact?

(1) $(4x^3 - y^3 + 2xy^2)dx + (2x^2y - 3xy^2 + 4y^3)dy = 0$

(2) $y^2 dx + \dfrac{x}{y} dy = 0$

(3) $(\dfrac{y^2}{2x^2} + \log_e y)dx + \dfrac{x^2 - y^2 + x}{xy} dy = 0$

(4) $(2x \log_e y + \dfrac{y^2}{x^2})dx + (\dfrac{x^2}{y} - 2y \log_e x)dy = 0$

(5) $(3x^2 + y + 1)dx + (3y^2 + x + 1)dy = 0$

****

66A

(1), (3) and (5) are exact.
(2) and (4) are not exact.

FRAME 67

Technique

Having established that the function $u(x,y)$ exists, its form can then be found by returning to equation (65.1),

i.e. $\quad \dfrac{\partial u}{\partial x} = P \ldots$ (i) $\quad$ and $\quad \dfrac{\partial u}{\partial y} = Q \ldots$ (ii),

and solving these simultaneously by integration.

For instance, starting from (i), we have

$$u = \int_{y\ \text{constant}} P(x,y)dx + \phi(y). \qquad (67.1)$$

Note that this integration w.r.t. x must be carried out with y held constant and that the constant of integration will be any arbitrary function ϕ of y. This is readily verified by partial differentiation since

$$\dfrac{\partial}{\partial x} \phi(y) = 0, \text{ for all } \phi(y).$$

The form of $\phi(y)$ is determined by substituting (67.1) into the other equation (ii).

FRAME 68

Example   (continued)

You have tested in FRAME 65 that the equation
$$(1 + 4xy + 2y^2)dx + (1 + 4xy + 2x^2)dy = 0$$
is exact.

We may therefore write
$$\frac{\partial u}{\partial x} = 1 + 4xy + 2y^2 \quad \ldots \ldots \text{ (i)}$$
$$\frac{\partial u}{\partial y} = 1 + 4xy + 2x^2 \quad \ldots \ldots \text{ (ii)}$$

From (i)
$$u = \int (1 + 4xy + 2y^2)dx + \phi(y)$$
y constant

i.e.
$$u = (x + 2x^2y + 2y^2x) + \phi(y). \qquad (68.1)$$

Substituting in (ii) we must have
$$\frac{\partial}{\partial y}\{(x + 2x^2y + 2y^2x) + \phi(y)\} = 1 + 4xy + 2x^2.$$

$$\therefore \quad 2x^2 + 4yx + \frac{d\phi}{dy} = 1 + 4xy + 2x^2$$

i.e.
$$\frac{d\phi}{dy} = 1.$$

Note: (a) Since ϕ depends on y only, ordinary differentiation is used.
 (b) All terms involving x should cancel at this stage, for the same reason.

Integrating $\frac{d\phi}{dy} = 1$,

we have $\quad \phi = y + B$

where B is an ordinary constant of integration.

Hence, (68.1) becomes
$$u = (x + 2x^2y + 2y^2x) + y + B.$$

The solution to the original exact d.e. is therefore
$$u = A \quad \text{or} \quad x + y + 2x^2y + 2y^2x = C. \qquad (C = A - B)$$

FIRST ORDER DIFFERENTIAL EQUATIONS

FRAME 69

In a similar manner, as an alternative method, we could have started from equation (ii), $\frac{\partial u}{\partial y} = Q$, of FRAME 67 to obtain

$$u = \int Q(x,y)dy + \psi(x) \qquad (69.1)$$
x constant

where $\psi(x)$ is any arbitrary function of x.

Substitute (69.1) in $\frac{\partial u}{\partial x} = P$ to find $\psi(x)$ and complete the solution.

****

69A

$$u = (y + 2xy^2 + 2x^2y) + \psi(x)$$

$$\frac{\partial u}{\partial x} = 2y^2 + 4xy + \frac{d\psi}{dx} = 1 + 4xy + 2y^2$$

$$\frac{d\psi}{dx} = 1 \quad yielding \quad \psi = x + B_1$$

and result follows as before.

FRAME 70

Now return to FRAME 66 and solve those equations which you have verified to be exact. You may care to solve some or all of these by the two methods given.

****

70A

(1) $\quad x^4 - xy^3 + x^2y^2 + y^4 = A$

(3) $\quad (x + 1) \log_e y - \frac{y^2}{2x} = A$

(5) $\quad x^3 + y^3 + xy + x + y = A$

FRAME 71

Equations reducible to exact by the use of integrating factors

An equation $Pdx + Qdy = 0$ which is not exact as it stands may sometimes be made exact by multiplying by some function of x and y, i.e. an integrating factor.

FRAME 71 continued

Example

Show that e^x is an integrating factor for the equation
$$(x + y)dx + dy = 0$$
and hence obtain the general solution.

In this equation $P = x + y$ and $Q = 1$

$$\therefore \quad \frac{\partial P}{\partial y} = 1 \quad \text{and} \quad \frac{\partial Q}{\partial x} = 0.$$

$\therefore$ It is not exact.

Multiplying by e^x we have
$$e^x(x + y)dx + e^x dy = 0$$
for which
$$P = e^x(x + y) \quad \text{and} \quad Q = e^x.$$

Check that the equation is now exact and complete the solution.

****

71A

$$u = \int e^x dy + \psi(x) \quad \textit{choosing the simpler Q function.}$$
$$x \textit{ constant}$$

$$\frac{d\psi}{dx} = xe^x$$

$$(x + y - 1)e^x = A.$$

FRAME 72

If you go back and look at any of the linear equations in FRAMES 50 - 63, you will see that the effect of multiplying such an equation throughout by the integrating factor was to make the L.H.S. an exact differential, i.e. $\frac{d}{dx}(uy)$. In that case we showed you how to find the integrating factor, but in other types of equation it is a more difficult problem and we shall not consider it here.

FIRST ORDER DIFFERENTIAL EQUATIONS

FRAME 73

Application

The technique of solving exact differential equations is often used more in integrating exact differentials than in actually solving differential equations. We shall take an example of such a problem.

Example

In two dimensional fluid motion, ϕ is the velocity potential and ψ the stream function at any point (x,y). They are connected by the equations

$$\frac{\partial \phi}{\partial x} = \frac{\partial \psi}{\partial y} \quad \text{and} \quad \frac{\partial \phi}{\partial y} = -\frac{\partial \psi}{\partial x}.$$

Find the equipotential lines (i.e. curves for which ϕ is constant) in the case of a sink of strength m at the origin, for which the stream function is

$$\psi = m \tan^{-1}(y/x).$$

$$\frac{\partial \psi}{\partial x} = m \frac{1}{1 + y^2/x^2} \cdot \frac{-y}{x^2} = -\frac{my}{x^2 + y^2}$$

$$\frac{\partial \psi}{\partial y} = \frac{mx}{x^2 + y^2}$$

$$\therefore \frac{\partial \phi}{\partial x} = \frac{mx}{x^2 + y^2} \quad \text{and} \quad \frac{\partial \phi}{\partial y} = \frac{my}{x^2 + y^2}$$

$$d\phi = \frac{mx}{x^2 + y^2} dx + \frac{my}{x^2 + y^2} dy.$$

You might like to check that the R.H.S. is an exact differential.

****

73A

If $P = \dfrac{mx}{x^2 + y^2}$ and $Q = \dfrac{my}{x^2 + y^2}$

$$\frac{\partial P}{\partial y} = -\frac{2mxy}{(x^2 + y^2)^2} = \frac{\partial Q}{\partial x}.$$

FRAME 74

$$\phi = \int Pdx + f(y)$$
$$\text{y constant}$$

$$= \int \frac{mx}{x^2 + y^2} \, dx + f(y)$$

$$= \frac{m}{2} \log_e(x^2 + y^2) + f(y)$$

$$\frac{\partial \phi}{\partial y} = \frac{my}{x^2 + y^2} + f'(y)$$

Comparing with Q, $f'(y) = 0$
$$f(y) = A$$

$$\therefore \quad \phi = \frac{m}{2} \log_e(x^2 + y^2) + A.$$

The constant A can be disregarded because it can be absorbed into the R.H.S. of the equation ϕ = constant.

Equipotential lines have equations of the form

$$\frac{m}{2} \log_e(x^2 + y^2) = c$$

i.e. $\quad x^2 + y^2 = C.$

FRAME 75

You have by now seen the origin of first order d.e.'s, their classification into various types, the techniques of solution and some applications.

In the following frame, a collection of miscellaneous examples relating to first order equations and their applications is given. These are not in any particular order and therefore you will require first of all to recognise and classify them before solution.

Do as many of these examples as you can, but you need not necessarily adhere to the order in which they are given!

Answers to these examples are supplied in FRAME 77. In some cases, hints have been provided.

FIRST ORDER DIFFERENTIAL EQUATIONS

FRAME 76

Miscellaneous Examples

Solve the following d.e.'s, subject to the conditions given where appropriate. Some examples from practical applications are included in this list and formulation of the d.e.'s may be required in some of these cases.

(1) $(4 - x^2)\dfrac{dy}{dx} + (2y + 1) = 0$, if $y = 0$ when $x = 1$.

(2) $(x^3 + y^3)\dfrac{dy}{dx} = x^2 y$, given that $y = 1$ when $x = 0$.

(3) $\dfrac{dy}{dx} = (x - y)^2$. (Hint: use $z = x - y$.)

(4) $\{y(1 + \dfrac{1}{x}) + \cos y\}dx + \{x + \log_e x - x \sin y\}dy = 0$, given that $y = \pi/2$ when $x = 1$.

(5) $(1 + x)\dfrac{dy}{dx} + (1 + 2x)y = (1 + x)^2$, given that $y = 0$ when $x = 1$.

(6) $\dfrac{dy}{dx} = \dfrac{2x + 9y - 20}{6x + 2y - 10}$, if $y = 2$ when $x = 2$.

(7) $(x + 1)\dfrac{dy}{dx} + (x + 2)y = 2 \sin x$, if $y = 0$ when $x = \dfrac{\pi}{4}$.

(8) $\dfrac{dy}{dx} + 1 = x - y + xy$, given that $y = 1$ when $x = 2$.

(9) $\dfrac{dy}{dx} = \dfrac{x + y + 3}{2x + 2y + 9}$.

(10) $\dfrac{dy}{dx} + 2y \tan x = \sin x$.

Some examples from practical applications follow and formulation of the d.e.'s may be required in some of them. We suggest that you select those which are most relevant for you.

(11) A body of mass m is projected with speed V on a rough table, coefficient of friction = μ, assumed constant. If the air resistance is proportional to the speed v (equal to mkv, say), find the distance travelled before the body comes to rest.

FIRST ORDER DIFFERENTIAL EQUATIONS

FRAME 76 continued

(12) A coil of inductance 3 H and resistance 12 Ω is connected to an applied voltage E(t). Find the current flowing t s after the circuit is closed in the following cases:

  (i) E(t) = 12 (12 V battery)

  (ii) E(t) = 240√2 sin 100πt (240 V, 50 cycle, mains).

(13) A bimolecular reaction is governed by the equation $\frac{dx}{dt} = k(5 - x)^2$, where x is the change in concentration at time t. x is zero initially and is found to have the value x = 1 when t = 10. Find the value of the reaction rate constant k and the values of x when t = 20 and t = 100. What is the ultimate value of x?

(14) In a thick cylinder subjected to internal pressure, the radial stress, P, at distance r from the axis of the cylinder is given by

$$P + r\frac{dP}{dr} = 2a - P$$

where a is a constant.

If the stress has magnitude P_o at the inner wall ($r = r_o$) and may be neglected at the outer wall ($r = r_1$) show that

$$P = -\frac{P_o r_o^2}{r_1^2 - r_o^2}\left(\frac{r_1^2}{r^2} - 1\right).$$

(15) A body takes 10 minutes to cool from 100°C to 80°C when the temperature of the surroundings is 20°C. How long does it take to cool from 80°C to 60°C? Assume that the rate of decrease of temperature is proportional to the excess of the temperature of the body over that of its surroundings.

(16) The mass of a rocket at time t is M and its speed is v. M consists of partly explosive material which is assumed to burn at a constant rate $m = -\frac{dM}{dt}$ and which is expelled backwards at a constant speed u relative to the rocket. Neglecting air resistance, the equation of motion is therefore

FIRST ORDER DIFFERENTIAL EQUATIONS

FRAME 76 continued

$$\frac{d}{dt}(Mv) - m(u - v) = -Mg.$$

Show that this reduces to

$$\frac{dv}{dt} = \frac{mu}{M_o - mt} - g$$

with $M(t) = M_o - mt$, M_o being the initial mass of the rocket and hence find the speed acquired in time T.

(17) A chain is coiled up near the edge of a smooth table and begins to fall over the edge. When a length x has fallen the equation of motion is

$$\frac{d}{dt}(mxv) = mxg$$

where m is the constant mass per unit length. Show that this reduces to

$$xv\frac{dv}{dx} + v^2 = gx$$

and, by using the substitution $y = v^2$, show that $v = \sqrt{\frac{2}{3}gx}$.

(18) The reaction $2NO + O_2 \rightarrow 2NO_2$ is governed by the equation

$$\frac{dx}{dt} = k(1 - 2x)^2(1 - x),$$

in which k is a constant, and x is the decrease in concentration of O_2 at time t, the initial concentrations of O_2 and NO being both equal to unity. When $t = 10$ the value of x is measured and found to be $1/4$. Evaluate the reaction rate constant k.

(19) A radio-active substance R_1 decays into a substance R_2 which is itself radio-active. At time t, N_1 atoms of R_1 and N_2 atoms of R_2 are present and are related by the equations

$$\frac{dN_1}{dt} = -\lambda_1 N_1$$

and $$\frac{dN_2}{dt} = \lambda_1 N_1 - \lambda_2 N_2,$$

subject to the conditions that $N_1 = N_o$ and $N_2 = 0$ initially. Solve the first equation to obtain N_1 as a function of t and use this result in the second equation to express N_2 similarly.

Answers to Miscellaneous Examples

(1) Separable $(2y + 1)^2 = \dfrac{3(2 - x)}{2 + x}$

(2) Homogeneous $y = vx$ gives

$$\dfrac{dx}{x} = -\dfrac{v^3 + 1}{v^4} dv.$$

$$y^3 = e^{x^3/y^3}$$

(3) Result is $1 - \dfrac{dz}{dx} = z^2$ which is separable.

$$\dfrac{1 + x - y}{1 - x + y} = Ae^{2x}$$

(4) Exact $y(x + \log_e x) + x \cos y = \pi/2$

(5) Linear I.F. $= \dfrac{e^{2x}}{1 + x}$

$$y = \tfrac{1}{2}(1 + x)(1 - e^{2-2x})$$

(6) $Y = 2x + 9y - 20$ and $X = 6x + 2y - 10$ produces the homogeneous equation $\dfrac{dY}{dX} = \dfrac{2X + 9Y}{6X + 2Y}$

$$\dfrac{dX}{X} = \left(\dfrac{2}{2 - V} + \dfrac{2}{1 + 2V}\right) dV \qquad (Y = VX)$$

$$(2x - y)^2 = 4(x + 2y - 5)$$

(7) Linear I.F. $= e^x(x + 1)$

$$y = \dfrac{\sin x - \cos x}{x + 1}$$

(8) Separable $\dfrac{dy}{y + 1} = (x - 1) dx$

$$y = 2e^{\tfrac{1}{2}x^2 - x} - 1$$

FIRST ORDER DIFFERENTIAL EQUATIONS

FRAME 77 continued

(9) $Y = x + y + 3$ gives $\dfrac{dY}{dx} - 1 = \dfrac{Y}{2Y + 3}$ which is separable.

$x + y + 4 = Ae^{x-2y}$

(10) Linear I.F. = $\sec^2 x$

$y = \cos x - A\cos^2 x$

(11) $mv\dfrac{dv}{dx} = -\mu mg - mkv$ Separable

Distance $= \displaystyle\int_0^V \dfrac{v\,dv}{kv + \mu g}$

$= \dfrac{1}{k}\{V - \dfrac{\mu g}{k}\log_e(1 + \dfrac{kV}{\mu g})\}$

(12) $L\dfrac{di}{dt} + Ri = E(t)$ Linear

$L = 3$, $R = 12$ and (i) $E = 12$, (ii) $E = 240\sqrt{2}\sin 100\pi t$

(i) $i = (1 - e^{-4t})$ A

(ii) $i = \dfrac{20\sqrt{2}}{1 + 625\pi^2}(\sin 100\pi t - 25\pi\cos 100\pi t + 25\pi e^{-4t})$ A

(13) Separable $kt = \dfrac{1}{5 - x} - \dfrac{1}{5}$

$k = 1/200$

$x = \dfrac{5}{3}$ when $t = 20$

$x = \dfrac{25}{7}$ when $t = 100$

$x \to 5$ as $t \to \infty$

(14) Linear I.F. = r^2

$r^2 P = ar^2 + A$

with $P = -P_o$ when $r = r_o$

and $P = 0$ when $r = r_1$

FRAME 77 continued

(15) $\dfrac{d\theta}{dt} = -k(\theta - \theta_o)$ Separable

θ = body temperature, θ_o = temperature of surroundings

Solution is $\theta - \theta_o = Ae^{-kt}$,

with $\theta = 100$ at $t = 0$, $\theta = 80$ at $t = 10$ and $\theta_o = 20$

$t = \dfrac{10 \log_e 2}{\log_e 4/3} = 24 \cdot 1$ when $\theta = 60$.

Time taken = $14 \cdot 1$ minutes.

(16) Speed = $mu \displaystyle\int_o^T \dfrac{dt}{M_o - mt} - gT$

$= u \log_e \dfrac{M_o}{M_o - mT} - gT$

(17) $y = v^2$ gives $x\dfrac{dy}{dx} + 2y = 2gx$ which is linear with

I.F. x^2 and subject to $v = 0$ when $x = 0$.

(18) Separable $\log_e \dfrac{1 - 2x}{1 - x} + \dfrac{1}{1 - 2x} = kt + 1$

$k = \dfrac{1}{10}(1 - \log_e 3/2) = \cdot 0595$

(19) $\dfrac{dN_1}{dt} = -\lambda_1 N_1$ Separable

$N_1 = N_o e^{-\lambda_1 t}$

$\therefore \dfrac{dN_2}{dt} + \lambda_2 N_2 = \lambda_1 N_o e^{-\lambda_1 t}$

This is linear with I.F. $e^{\lambda_2 t}$.

$N_2 = \dfrac{\lambda_1 N_o}{\lambda_1 - \lambda_2}(e^{-\lambda_2 t} - e^{-\lambda_1 t})$.

SECOND ORDER DIFFERENTIAL EQUATIONS WITH CONSTANT COEFFICIENTS

– SOLUTION BY TRIAL METHODS

A PROGRAMMED TEXT

A. C. Bajpai
J. Hyslop

INSTRUCTIONS

This programme constitutes a self-instructional course on the solution of second order differential equations with constant coefficients by the method of trial solution.

The programme is divided up into a number of FRAMES which are to be worked *in the order given*. You will be required to participate in many of these frames and in such cases the answers are provided in ANSWER FRAMES, designated by the letter A following the frame number. Steps in the working are given where this is considered helpful. The answer frame is separated from the main frame by a line of asterisks: ******************. Keep the answers covered until you have written your own response. If your answer is wrong, go back and try to see why. Do not proceed to the next frame until you have corrected any mistakes in your attempt and are satisfied that you understand the contents up to this point.

Before reading this programme it is necessary that you are familiar with the following

Prerequisites

Differentiation of standard functions and products.

Complex numbers — algebraic manipulation, equating real and imaginary parts, exponential form.

C O N T E N T S

Instructions

FRAMES

1 - 5	Introduction
6	Definition
7 - 17	Forms of Solution
18 - 31	Complementary Function
32 - 44	Particular Integral; cases (i) - (iv)
45 - 48	Particular Integral; sum and product forms
49 - 54	Particular Integral; cases of failure
55 - 57	Boundary Conditions
58 - 65	Applications
66	Summary
67	Miscellaneous Examples
68	Answers to Miscellaneous Examples

SECOND ORDER DIFFERENTIAL EQUATIONS

FRAME 1

Introduction

A detailed study of the complete subject of Second Order Differential Equations would involve a considerable amount of time and effort in discussing the theory and methods of solution of the many types of equation arising. However, one particular type occurs much more frequently than any other when considering applications arising from science and engineering. This type is the simplest of all the second order equations and is called the constant coefficient type, the general form being

$$a \frac{d^2y}{dx^2} + b \frac{dy}{dx} + cy = Q,$$

in which a, b and c are the constant coefficients, and Q is a function of the variable x.

FRAME 2

Many instances of equations of this type may be cited from science and engineering. For example, when considering the problem of the bending of a light beam of length ℓ under the action of its own weight and a compressive load P at its ends, which are clamped horizontally, we find that the displacement y at position x is given by the differential equation

$$EI \frac{d^2y}{dx^2} + Py = G - \tfrac{1}{2}w\ell x + \tfrac{1}{2}wx^2$$

where w is the weight per unit length, EI is the constant flexural rigidity and G is the magnitude of the clamping couple.

FRAME 3

Again, an electrical circuit consisting of an inductance L H, a resistance R Ω and a condenser of capacitance C F all connected in series to a generator supplying an e.m.f. E V, which may vary with time, gives rise to the equation

$$L \frac{d^2q}{dt^2} + R \frac{dq}{dt} + \frac{q}{C} = E,$$

specifying the charge q on the condenser at time t.

FRAME 4

Finally, in an oscillating mechanical system which is subject to a restoring force proportional to the displacement x from the equilibrium position and to a frictional resistance proportional to the speed, we find that, if the support is also made to vibrate sinusoidally, the equation of motion may be written as

$$\frac{d^2x}{dt^2} + 2k\frac{dx}{dt} + \omega^2 x = a \sin nt$$

where k, ω and a are constants.

FRAME 5

These three examples will be considered more fully later in the programme. They are all instances of the type of equation in FRAME 1 and it is the purpose of this programme to develop a suitable method of solution.

The method of solution adopted here is the METHOD OF TRIAL SOLUTIONS. This is the most elementary of the methods available involving only a knowledge of simple differentiation. In most text books it tends to be dismissed after a brief mention and the more sophisticated techniques such as D-operator or Laplace Transforms are considered in detail. Whilst it is intended to treat these more advanced methods in later programmes, the authors consider that a study of the simpler trial method is also very rewarding. Its main virtue is simplicity and, in addition, its directness contributes to a greater understanding of the structure of the solutions of the equations involved. This is valuable in that the ideas introduced are useful when considering extensions to more complicated differential equations which involve the use of techniques such as variation of parameters, which will also be discussed in a later programme.

SECOND ORDER DIFFERENTIAL EQUATIONS

FRAME 6

Definition

As mentioned in FRAME 1 the linear second order differential equation with constant coefficients may be written in the form

$$a \frac{d^2y}{dx^2} + b \frac{dy}{dx} + cy = Q, \qquad (6.1)$$

where a, b and c are constants and Q is a function of x only.

Which of the following differential equations (d.e.'s in future) are of the above type?

(1) $\quad 2 \dfrac{d^2y}{dx^2} + \dfrac{dy}{dx} - y = 3e^{2x}$

(2) $\quad \dfrac{d^2y}{dx^2} - 4 \dfrac{dy}{dx} + 4y = 0$

(3) $\quad \dfrac{d^2y}{dx^2} + x \dfrac{dy}{dx} - 2y = \sin x$

(4) $\quad \dfrac{d^2y}{dx^2} = - n^2 y$

(5) $\quad \dfrac{d^2y}{dx^2} - 8 \dfrac{dy}{dx} + 25y = e^{-x} \cos x$

(6) $\quad \dfrac{d^2x}{dt^2} - 3 \dfrac{dx}{dt} + 2x = \tfrac{1}{2} t$

****

6A

(1) *Yes.*

(2) *Yes. Q = 0 here.*

(3) *No. The coefficient of $\frac{dy}{dx}$ is not constant.*

(4) *Yes. a = 1, b = 0, c = n^2 and Q = 0.*

(5) *Yes.*

(6) *Yes. Note that the variables are x and t.*

FRAME 7

Forms of Solution

The solution of the d.e. (6.1) is closely related to that of the simpler equation

$$a \frac{d^2y}{dx^2} + b \frac{dy}{dx} + cy = 0. \qquad (7.1)$$

In this equation, known as the reduced equation, Q is equal to zero.

What is the reduced equation in example (1) of FRAME 6?

****

7A

$$2 \frac{d^2y}{dx^2} + \frac{dy}{dx} - y = 0 \qquad (7A.1)$$

FRAME 8

Let us now consider the solution of the reduced equation (7.1). The solution of the simpler first order equation $\frac{dy}{dx} + ky = 0$ is $y = Ae^{-kx}$ (see FRAME 26 in the programme "First Order Differential Equations" in this series). The exponential form of solution is not really surprising as the exponential function is the only function whose differential coefficient is a constant multiple of itself.

Extending this idea, it is reasonable to expect that equation (7.1) might have an exponential solution as the relationship between y, $\frac{dy}{dx}$ and $\frac{d^2y}{dx^2}$ is linear, and the second derivative of an exponential function is also a multiple of itself, i.e. if $y = e^{mx}$, where m is a constant,

then $\frac{dy}{dx} = me^{mx} = my$

and $\frac{d^2y}{dx^2} = m^2 e^{mx} = m^2 y.$

This suggests the possibility of using $y = e^{mx}$ as a trial solution of the reduced equation.

Now write down the equation obtained on substituting this trial solution into the reduced equation (7A.1)

****

SECOND ORDER DIFFERENTIAL EQUATIONS

8A

$$(2m^2 + m - 1)e^{mx} = 0$$
$$\text{i.e.} \quad 2m^2 + m - 1 = 0, \qquad (8A.1)$$
$$\text{since} \quad e^{mx} \neq 0.$$

FRAME 9

(8A.1) is a quadratic equation in m, which is referred to as the AUXILIARY EQUATION (A.E.). Solve it and obtain the corresponding solutions of the reduced equation.

****

9A

$$(2m - 1)(m + 1) = 0$$
$$\therefore m = \tfrac{1}{2} \text{ or } m = -1.$$

The corresponding solutions of the d.e. in 7A are $y = e^{\frac{1}{2}x}$ *and* $y = e^{-x}$.

FRAME 10

You will note that <u>two</u> independent solutions have been obtained, since the auxiliary equation, which is a quadratic, has two distinct roots. Verify, by direct substitution in the reduced equation, that $y = e^{\frac{1}{2}x}$ and $y = e^{-x}$ separately satisfy it.

****

10A

$$y = e^{\frac{1}{2}x}; \qquad L.H.S. = 2 \cdot \tfrac{1}{4} e^{\frac{1}{2}x} + \tfrac{1}{2} e^{\frac{1}{2}x} - e^{\frac{1}{2}x} = 0.$$

$$y = e^{-x}; \qquad L.H.S. = 2e^{-x} - e^{-x} - e^{-x} = 0.$$

FRAME 11

Now verify that $y = Ae^{\frac{1}{2}x}$ and $y = Be^{-x}$ and also the sum $y = Ae^{\frac{1}{2}x} + Be^{-x}$ satisfy the reduced equation (7A.1), A and B being arbitrary constants.

****

11A

$$y = Ae^{\frac{1}{2}x} \quad ; \quad A(2 \cdot \frac{1}{4} + \frac{1}{2} - 1)e^{\frac{1}{2}x} = 0.$$

$$y = Be^{-x} \quad ; \quad B(2 \cdot 1 - 1 - 1)e^{-x} = 0.$$

$$y = Ae^{\frac{1}{2}x} + Be^{-x} \; ; \quad A(2 \cdot \frac{1}{4} + \frac{1}{2} - 1)e^{\frac{1}{2}x} + B(2 \cdot 1 - 1 - 1)e^{-x} = 0.$$

FRAME 12

The solution $y = Ae^{\frac{1}{2}x} + Be^{-x}$ to the reduced equation

$$2\frac{d^2y}{dx^2} + \frac{dy}{dx} - y = 0$$

is referred to as the COMPLEMENTARY FUNCTION (C.F.). It contains the required number of arbitrary constants: two for a second order d.e. (refer to FRAMES 6 - 8 of the programme "First Order Differential Equations" in this series).

FRAME 13

If we return to the complete d.e. of example (1) in FRAME 6,

$$2\frac{d^2y}{dx^2} + \frac{dy}{dx} - y = 3e^{2x},$$

it is obvious that the C.F. is not the complete solution. An extra term needs to be added to balance the $3e^{2x}$ on the R.H.S. In view of the remarks made in FRAME 8, to obtain this extra term, the <u>trial solution</u> $y = \alpha e^{2x}$ is suggested. Determine α by direct substitution in the above d.e.

****

13A

$$(2 \cdot 4\alpha + 2\alpha - \alpha)e^{2x} \equiv 3e^{2x}$$

$$\therefore \alpha = \frac{1}{3}.$$

$\therefore$ *The required additional term is* $\frac{1}{3}e^{2x}$.

SECOND ORDER DIFFERENTIAL EQUATIONS

FRAME 14

Now write down the complete solution of the d.e. and verify by substitution that it does, in fact, satisfy the d.e.

****

14A

$$y = Ae^{\frac{1}{2}x} + Be^{-x} + \frac{1}{3}e^{2x}$$

$$\frac{dy}{dx} = y' = \frac{1}{2}Ae^{\frac{1}{2}x} - Be^{-x} + \frac{2}{3}e^{2x}$$

$$\frac{d^2y}{dx^2} = y'' = \frac{1}{4}Ae^{\frac{1}{2}x} + Be^{-x} + \frac{4}{3}e^{2x}$$

Whence $2y'' + y' - y = 3e^{2x} = $ R.H.S.

FRAME 15

The complete solution

$$y = (Ae^{\frac{1}{2}x} + Be^{-x}) + \frac{1}{3}e^{2x}$$

of the d.e. obtained in 14A is referred to as the GENERAL SOLUTION (G.S.). You will observe that it consists of two parts, the C.F. $(Ae^{\frac{1}{2}x} + Be^{-x})$ and the extra trial solution $\frac{1}{3}e^{2x}$. Observe also that the latter term contains no arbitrary constant since in $y = \alpha e^{3x}$, α must be assigned the value $\frac{1}{3}$ to ensure that the d.e. is satisfied. For this reason this part of the G.S. is referred to as the PARTICULAR INTEGRAL (P.I.). In short

$$\text{G.S.} = \text{C.F.} + \text{P.I.}$$

The C.F. contains two arbitrary constants. The P.I. contains no arbitrary constant.

It should be noted that this addition of partial solutions to produce a complete or general solution is justified only because of the linear nature of the d.e.

FRAME 16

In general, the complete solution of the equation (6.1)
$$ay'' + by' + cy = Q$$
may be written as
$$y = (Au + Bv) + \phi \qquad (16.1)$$
where $y = u$ and $y = v$ are independent solutions of the reduced equation
$$ay'' + by' + cy = 0$$
and $y = \phi$ is a particular solution of equation (6.1).

$(Au + Bv)$ is the C.F. and ϕ is the P.I. giving the G.S. as in (16.1).

You may care to confirm this by direct substitution as in the previous example.

****

16A

Since $y = u$ and $y = v$ are solutions of the reduced equation we have
$$au'' + bu' + cu = 0$$
$$av'' + bv' + cv = 0.$$

Also, since $y = \phi$ is a solution of the complete d.e. we have
$$a\phi'' + b\phi' + c\phi = Q.$$

Multiplying the first equation by A, the second equation by B and adding the sum to the third equation we have, on adding vertically
$$a(Au'' + Bv'' + \phi'') + b(Au' + Bv' + \phi') + c(Au + Bv + \phi) = Q$$

showing clearly that $y = Au + Bv + \phi$

is a solution of $ay'' + by' + cy = Q$.

SECOND ORDER DIFFERENTIAL EQUATIONS

FRAME 17

In the following frames we shall discuss methods of obtaining the general solutions of this type of d.e. The discussion will be divided into two parts:
(i) the complementary function and (ii) the particular integral.

FRAME 18

The Complementary Function (C.F.)

As suggested in FRAME 8 we try $y = e^{mx}$ as a solution of the reduced equation
$$ay'' + by' + cy = 0.$$
The resulting auxiliary equation is the quadratic
$$am^2 + bm + c = 0.$$
The two independent solutions giving the C.F. depend on the two roots m_1 and m_2, say, of this quadratic. You will recall that three cases arise:

(i)   real unequal roots, when $b^2 > 4ac$
(ii) real equal roots, when $b^2 = 4ac$
(iii) complex conjugate roots, when $b^2 < 4ac$.

We shall discuss the form of the C.F. corresponding to each of these three cases in turn.

FRAME 19

Case (i)

The roots are the real numbers m_1 and m_2 ($m_1 \neq m_2$) and the corresponding C.F. is
$$y = (Ae^{m_1 x} + Be^{m_2 x}).$$
For instance, in example (1) of FRAME 6, the reduced equation
$$2\frac{d^2 y}{dx^2} + \frac{dy}{dx} - y = 0$$
was treated in FRAMES 8 - 11 where it was shown that $m_1 = \tfrac{1}{2}$ and $m_2 = -1$ and the C.F. was $y = Ae^{\tfrac{1}{2}x} + Be^{-x}$.

FRAME 19 continued

Now write down the auxiliary equation in each of the following cases and hence obtain the corresponding C.F.'s.

(1) $y'' - 3y' + 2y = 0$

(2) $2y'' + 5y' - 3y = 0$

(3) $\dfrac{d^2x}{dt^2} + 2\dfrac{dx}{dt} - 3x = 0$

(4) $y'' - 4y = 0$

(5) $y'' + 3y' = 0$

****

19A

(1) A.E. $m^2 - 3m + 2 = 0$; Roots 1, 2; C.F. $y = Ae^x + Be^{2x}$

(2) A.E. $2m^2 + 5m - 3 = 0$; Roots $\frac{1}{2}$, -3; C.F. $y = Ae^{\frac{1}{2}x} + Be^{-3x}$

(3) A.E. $m^2 + 2m - 3 = 0$; Roots 1, -3; C.F. $x = Ae^t + Be^{-3t}$

(4) A.E. $m^2 - 4 = 0$; Roots 2, -2; C.F. $y = Ae^{2x} + Be^{-2x}$

(5) A.E. $m^2 + 3m = 0$; Roots 0, -3; C.F. $y = A + Be^{-3x}$

(Note that $Ae^{0 \cdot x} = A$)

FRAME 20

Case (ii)

In this case $m_1 = m_2 = n$, say, where n is real. As in the previous case, you might consider that the C.F. would be $y = Ae^{nx} + Be^{nx}$. However, this could be written as $y = (A + B)e^{nx} = Ce^{nx}$ where C ($= A + B$) is a single arbitrary constant. This means that the solution contains only one arbitrary constant instead of the required two, showing that the C.F. is incomplete. It is therefore necessary to use some additional device to obtain the complete C.F.

We shall show that the need is met by using a trial solution $y = e^{nx}f$, where f is a function of x to be determined. Note that $n = -b/2a$ here since $an^2 + bn + c = 0$ with $b^2 = 4ac$ for equal roots of the A.E.

SECOND ORDER DIFFERENTIAL EQUATIONS

FRAME 20 continued

Differentiation yields

$$y' = e^{nx}(f' + nf)$$

and $$y'' = e^{nx}(f'' + 2nf' + n^2 f).$$

Substitute these into the d.e. $ay'' + by' + cy = 0$ and show that it reduces to

$$f'' = 0.$$

****

20A

$$e^{nx}\{a(f'' + 2nf' + n^2 f) + b(f' + nf) + cf\} = 0$$

$$\therefore \quad af'' + (2an + b)f' + (an^2 + bn + c)f = 0.$$

This immediately reduces to $f'' = 0$ since the coefficients of f' and f vanish.

FRAME 21

To obtain f from the simple d.e. $f'' = 0$ or $\dfrac{d^2 f}{dx^2} = 0$, we integrate twice.

Thus, $$\dfrac{df}{dx} = B$$

and $\therefore$ $$f = A + Bx$$

where A and B are arbitrary constants of integration. This shows that the complete C.F. is

$$y = (A + Bx)e^{nx}.$$

You may care to confirm this result by direct substitution in the d.e. $ay'' + by' + cy = 0$.

****

21A

$$y = e^{nx}(A + Bx)$$

$$y' = e^{nx}\{n(A + Bx) + B\}$$

$$y'' = e^{nx}\{n^2(A + Bx) + 2nB\}$$

21A continued

Hence,
$$ay'' + by' + cy = e^{nx}\left[a\{n^2(A + Bx) + 2nB\} + b\{n(A + Bx) + B\} + c(A + Bx)\right]$$
$$= e^{nx}\{(A + Bx)(an^2 + bn + c) + B(2an + b)\}$$
$$= 0$$

since $n = -\dfrac{b}{2a}$ and $an^2 + bn + c = 0$.

FRAME 22

Note that the part of the solution which would have been repeated because of the equal roots has been modified by multiplying it by x. This device will again be used later in obtaining particular integrals in certain cases. (See FRAME 49 et seq.)

FRAME 23

For instance the A.E. for $y'' - 8y' + 16y = 0$ is $m^2 - 8m + 16 = 0$ with the repeated root $m = 4$, and the C.F. is $y = (A + Bx)e^{4x}$.

Now solve the following:

(1) $\quad y'' - 6y' + 9y = 0$

(2) $\quad 4y'' - 12y' + 9y = 0$

(3) $\quad y'' + 2y' + y = 0$

****

23A

(1) $\quad y = (A + Bx)e^{3x}$

(2) $\quad y = (A + Bx)e^{3x/2}$

(3) $\quad y = (A + Bx)e^{-x}$

SECOND ORDER DIFFERENTIAL EQUATIONS

FRAME 24

Case (iii)

In this case, in the A.E. $am^2 + bm + c = 0$, $b^2 < 4ac$ and the roots are

$$\frac{-b \pm \sqrt{b^2 - 4ac}}{2a} = p \pm iq, \quad \text{say},$$

where $p = -b/2a$ and $q = \sqrt{4ac - b^2}/2a$.

The C.F. may therefore be written as

$$y = A_1 e^{(p+iq)x} + B_1 e^{(p-iq)x} \qquad (24.1)$$

where A_1 and B_1 are arbitrary constants.

FRAME 25

The form of solution given in (24.1) may be written more conveniently in the form

$$y = e^{px}\{A_1 e^{iqx} + B_1 e^{-iqx}\}$$

$$= e^{px}\{A_1(\cos qx + i \sin qx) + B_1(\cos qx - i \sin qx)\}$$

i.e. $\quad y = e^{px}(A \cos qx + B \sin qx) \qquad (25.1)$

where $A = A_1 + B_1$ and $B = i(A_1 - B_1)$ are new arbitrary constants. Note that A and B are completely arbitrary and need not involve imaginary quantities; nor need A_1 and B_1 be purely real.

FRAME 26

By suitably redefining the constants, the standard result (25.1) may be written in alternative forms which are more suitable in certain applications. For instance, show that (25.1) may be written in the forms

$$y = Ce^{px} \sin(qx + \theta)$$

or $\quad y = Ce^{px} \cos(qx - \phi),$

by using the usual trigonometric processes to specify the arbitrary constants C, θ and ϕ.

****

26A

In (25.1) write $A = C \sin \theta$, $B = C \cos \theta$ giving $C = \sqrt{A^2 + B^2}$ and $\tan \theta = A/B$.

In the second case, write $A = C \cos \phi$
$\qquad\qquad\qquad\qquad\quad B = C \sin \phi$

FRAME 27

Now solve the following equations:

(1) $\quad y'' - 8y' + 25y = 0$

(2) $\quad y'' + 4y' + 5y = 0$

(3) $\quad \dfrac{d^2x}{dt^2} + \dfrac{dx}{dt} + x = 0$

(4) $\quad \dfrac{d^2x}{dt^2} = -\omega^2 x$

****

27A

(1) $\quad y = e^{4x}(A \cos 3x + B \sin 3x)$

(2) $\quad y = e^{-2x}(A \cos x + B \sin x)$

(3) $\quad x = e^{-t/2}\left(A \cos \dfrac{\sqrt{3}}{2}t + B \sin \dfrac{\sqrt{3}}{2}t\right)$

(4) $\quad x = A \cos \omega t + B \sin \omega t$.

(Note that $p = 0$ here, since A.E. is $m^2 = -\omega^2$.)

FRAME 28

You may find it useful to refer to the following summary for obtaining C.F.'s.

$\quad$ Case (i) $\quad$ Real unequal roots m_1 and m_2
$\qquad\qquad\qquad$ C.F. $y = A e^{m_1 x} + B e^{m_2 x}$

$\quad$ Case (ii) $\quad$ Real equal roots $m_1 = m_2 = n$
$\qquad\qquad\qquad$ C.F. $y = (A + Bx)e^{nx}$

$\quad$ Case (iii) $\quad$ Complex conjugate roots $p \pm iq$
$\qquad\qquad\qquad$ C.F. $y = e^{px}(A \cos qx + B \sin qx)$.

SECOND ORDER DIFFERENTIAL EQUATIONS

FRAME 29

To show how these cases arise in practice we consider the following physical application. A particle of unit mass is subjected to a restoring force $\omega^2 x$ directed towards the origin and to a resistance $2k\dot{x}$. ω and k are constants, x is the displacement at time t and dots denote differentiations with respect to t. The equation of motion is

$$\ddot{x} = -2k\dot{x} - \omega^2 x$$

for which the A.E. is $m^2 + 2km + \omega^2 = 0$ and its roots m_1 and m_2 are $-k \pm \sqrt{k^2 - \omega^2}$.

FRAME 30

Cases (i) - (iii) arise and are interpreted as follows:

<u>Case (i)</u>   The resistance is large enough for k to be greater than ω, i.e. $k^2 - \omega^2 > 0$ here. The solution is

$$x = A \exp\{(-k + \sqrt{k^2 - \omega^2})t\} + B \exp\{(-k - \sqrt{k^2 - \omega^2})t\}.$$

Both exponentials decay as $t \to \infty$. The solution is therefore non-oscillatory and the system is said to be over-damped. The graph of the displacement against time is as shown, there being three possibilities according to the initial conditions.

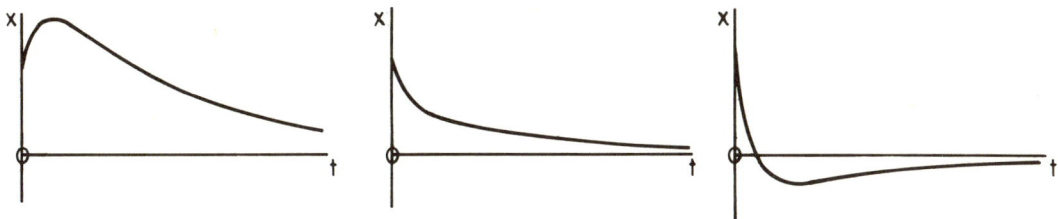

<u>Case (ii)</u>   The resistance term just balances the restoring term so that $k = \omega$, i.e. $k^2 - \omega^2 = 0$ here.
The solution is

$$x = e^{-kt}(A + Bt).$$

The solution is again non-oscillatory and $x \to 0$ as $t \to \infty$. The system is said to be critically damped. The displacement-time graph is similar to that for Case (i).

FRAME 30 continued

<u>Case (iii)</u>   The resistance is small and $k^2 - \omega^2 < 0$ here.

The solution is
$$x = e^{-kt}\{A \cos (\sqrt{\omega^2 - k^2}\, t) + B \sin (\sqrt{\omega^2 - k^2}\, t)\}$$

or $x = Ce^{-kt} \sin (\sqrt{\omega^2 - k^2}\, t + \theta)$, say,

as in FRAME 26. This represents an oscillation of frequency $(\omega^2 - k^2)^{\frac{1}{2}}/2\pi$ whose amplitude is <u>damped</u> by the factor e^{-kt} as represented in the diagram.

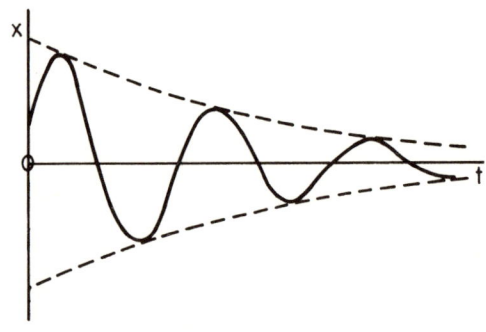

FRAME 31

For further practice, solve the following d.e.'s:

(1) $2y'' + 3y' - 2y = 0$

(2) $y'' - 2y' + 2y = 0$

(3) $y'' - 2y' + y = 0$

(4) $y'' = -4y$

(5) $y'' = 4y$

(6) $36y'' - 36y' + 13y = 0$

(7) $3y'' + 2y' = 0$

(8) $16y'' - 8y' + y = 0$

****

SECOND ORDER DIFFERENTIAL EQUATIONS

31A

(1) $y = Ae^{\frac{1}{2}x} + Be^{-2x}$

(2) $y = e^{x}(A \cos x + B \sin x)$

(3) $y = e^{x}(A + Bx)$

(4) $y = A \cos 2x + B \sin 2x$

(5) $y = Ae^{2x} + Be^{-2x}$

(6) $y = e^{\frac{1}{2}x}(A \cos \frac{1}{3}x + B \sin \frac{1}{3}x)$

(7) $y = A + Be^{-2x/3}$

(8) $y = e^{\frac{1}{4}x}(A + Bx)$.

FRAME 32

The Particular Integral (P.I.)

We return now to the solution of the complete d.e. (6.1)

$$ay'' + by' + cy = Q.$$

You have already seen how to obtain the C.F. corresponding to the reduced equation (7.1)

$$ay'' + by' + cy = 0$$

in which the R.H.S. is zero. You will recall from FRAMES 15 and 16 that to obtain the complete or general solution to (6.1) we still require to find the P.I.'s corresponding to various forms of Q. We shall restrict ourselves to the following forms for Q which occur most frequently in applications from science and engineering:

(i) $ke^{\lambda x}$
(ii)   low order polynomial expressions such as $\lambda x^2 + \mu x + \nu$
(iii) $k \cos \mu x$ or $k \sin \mu x$
(iv) $ke^{\lambda x} \cos \mu x$ or $ke^{\lambda x} \sin \mu x$

k, λ, μ and ν are given constants.

SECOND ORDER DIFFERENTIAL EQUATIONS

FRAME 33

Case (i) $Q = ke^{\lambda x}$

As in the example considered in FRAMES 13 and 13A, the trial solution for the P.I. should be $y = \alpha e^{\lambda x}$ and the value of α determined by direct substitution in the given d.e.

Find the general solution of the d.e.

$$y'' - 3y' + 2y = 2e^{-4x}$$

by first obtaining the C.F. and then the P.I. for $Q = 2e^{-4x}$.

****

33A

A.E. is $m^2 - 3m + 2 = 0$ with roots 1, 2.

C.F. is $Ae^x + Be^{2x}$.

P.I. for $Q = 2e^{-4x}$; try $y = \alpha e^{-4x}$.

Then $(16 + 12 + 2)\alpha e^{-4x} \equiv 2e^{-4x}$

$\therefore \quad \alpha = \dfrac{1}{15}$

P.I. is $\dfrac{1}{15} e^{-4x}$.

$\therefore$ G.S. is $y = Ae^x + Be^{2x} + \dfrac{1}{15} e^{-4x}$.

FRAME 34

Case (ii) $Q = \lambda x^2 + \mu x + \nu$

In this, a low order polynomial case, we confine our attention to the above quadratic form by way of illustration.

The suggested form of trial solution for the P.I. is

$$y = \alpha x^2 + \beta x + \gamma$$

where α, β and γ are to be determined.

Consider the d.e.

$$2y'' + 5y' - 3y = 2x^2 - x + 1.$$

SECOND ORDER DIFFERENTIAL EQUATIONS

FRAME 34 continued

The A.E. is $2m^2 + 5m - 3 = 0$ with roots $\tfrac{1}{2}$ and -3, so that the C.F. is $Ae^{\tfrac{1}{2}x} + Be^{-3x}$.

For the P.I. we try $y = \alpha x^2 + \beta x + \gamma$ as suggested above.

Substituting in the d.e. we find

$$-3(\alpha x^2 + \beta x + \gamma) + 5(2\alpha x + \beta) + 2(2\alpha) \equiv 2x^2 - x + 1,$$

noting that the terms have been written in reverse order to facilitate differentiation and substitution.

As indicated by the identity sign, this relation must be true for all values of x. Hence, we may equate the coefficients of powers of x, starting with the highest power:

coefficient of x^2	$-3\alpha = 2,$	giving $\alpha = -\tfrac{2}{3}$
coefficient of x	$-3\beta + 10\alpha = -1,$	giving $\beta = -\tfrac{17}{9}$
constant terms (i.e. coefficient of x^0)	$-3\gamma + 5\beta + 4\alpha = 1,$	giving $\gamma = -\tfrac{118}{27}$

therefore P.I. is $\quad y = -\tfrac{1}{27}(18x^2 + 51x + 118)$

and G.S. is $\quad y = Ae^{\tfrac{1}{2}x} + Be^{-3x} - \tfrac{1}{27}(18x^2 + 51x + 118).$

FRAME 35

Now consider the d.e.
$$2y'' + 5y' - 3y = Q.$$

Can you suggest the trial solution for the P.I. for each of the following forms of Q?

(1) $\quad x + 4$

(2) $\quad 9x^2$

(3) $\quad -3$

****

35A

(1) $y = \alpha x + \beta$. Note that no term in x^2 need be included since it does not arise in Q.
(2) $y = \alpha x^2 + \beta x + \gamma$. In this case terms in x and a constant must be included as well as the obvious αx^2 term, since they will arise on differentiation. If you are not convinced about this, see what happens if you use a trial solution $y = \alpha x^2$ in the next frame.
(3) $y = \alpha$. Terms in x and x^2 need not be included since they do not arise in Q.

FRAME 36

Solve the d.e. given in FRAME 35 for the three values of Q as specified in (1) (2) and (3).

****

36A

(1) $y = Ae^{\frac{1}{2}x} + Be^{-3x} - \frac{1}{9}(3x + 17)$.

(2) $y = Ae^{\frac{1}{2}x} + Be^{-3x} - \frac{1}{3}(9x^2 + 30x + 62)$. If you tried $y = \alpha x^2$ for the P.I. you will have obtained three inconsistent equations for α.

(3) $y = Ae^{\frac{1}{2}x} + Be^{-3x} + 1$.

The C.F. is, of course, the same for each Q.

FRAME 37

Case (iii) $Q = k \cos \mu x$ or $k \sin \mu x$

Taking $Q = k \cos \mu x$, for example, the trial solution is
$y = \alpha \cos \mu x + \beta \sin \mu x$ where α and β are to be determined. The sine term must be included since it arises on differentiation of the obvious $\alpha \cos \mu x$ term. Similar remarks apply in the case $Q = k \sin \mu x$. We shall illustrate this method by solving the d.e.

$$y'' + 3y' + 2y = 2 \cos 3x. \qquad (37.1)$$

The A.E. is $m^2 + 3m + 2 = 0$ with roots -1 and -2, giving the C.F. $Ae^{-x} + Be^{-2x}$.

SECOND ORDER DIFFERENTIAL EQUATIONS

FRAME 37 continued

For the P.I. we try $y = \alpha \cos 3x + \beta \sin 3x$.

Substitution yields
$$2(\alpha \cos 3x + \beta \sin 3x) + 3(-3\alpha \sin 3x + 3\beta \cos 3x)$$
$$+ (-9\alpha \cos 3x - 9\beta \sin 3x) \equiv 2 \cos 3x.$$

The coefficients of $\cos 3x$ and $\sin 3x$ may be equated giving
$$2\alpha + 9\beta - 9\alpha = 2$$
$$2\beta - 9\alpha - 9\beta = 0.$$

Solve these to obtain α and β and hence write down the general solution to the d.e.

***

****

37A

$$\alpha = -\frac{7}{65}, \qquad \beta = \frac{9}{65}.$$

G.S. is $y = Ae^{-x} + Be^{-2x} + \frac{1}{65}(9 \sin 3x - 7 \cos 3x).$

FRAME 38

Use a similar method to solve
$$y'' - 2y' + y = 5 \sin 2x \qquad (38.1)$$

****

38A

A.E. is $m^2 - 2m + 1 = 0$ *with repeated root* $m = 1$.
C.F. is $(A + Bx)e^x$.
For P.I. try $y = \alpha \sin 2x + \beta \cos 2x$.

Substitution in the d.e. and equating coefficients of sin 2x and cos 2x yields
$$-3\alpha + 4\beta = 5$$
$$-4\alpha - 3\beta = 0$$

Solving these we obtain $\alpha = -\frac{3}{5}$ *and* $\beta = \frac{4}{5}$.

Therefore G.S. is $y = (A + Bx)e^x + \frac{1}{5}(4 \cos 2x - 3 \sin 2x)$

FRAME 39

Alternative Method

In the case $Q = k \cos \mu x$ or $k \sin \mu x$ we may consider instead $Q = k e^{i\mu x}$ which contains $k \cos \mu x$ and $k \sin \mu x$ as its real (Re) and imaginary (Im) parts respectively. The trial solution $\alpha e^{i\mu x}$ is suggested as in Case (i). The value of α (possibly complex) is obtained by substitution and the required P.I. is then found by taking Re or Im parts as appropriate. By way of illustration we shall solve equation (37.1) by this method.

$$y'' + 3y' + 2y = 2 \cos 3x$$

C.F. is $Ae^{-x} + Be^{-2x}$ as before

$Q = 2 \cos 3x = \text{Re}(2e^{3ix})$.

The P.I. for this equation is therefore the real part of the P.I. of the equation

$$y'' + 3y' + 2y = 2e^{3ix},$$

for which we try $y = \alpha e^{3ix}$.

Substitution gives

$$\alpha e^{3ix}\{2 + 3.3i + (3i)^2\} \equiv 2e^{3ix},$$

from which $\alpha = \dfrac{2}{-7 + 9i}$

therefore the P.I. for $2e^{3ix} = \dfrac{2}{-7 + 9i} e^{3ix}$

and hence the P.I. for $2 \cos 3x = \text{Re}\{\dfrac{2}{-7 + 9i} e^{3ix}\}$.

Simplify by the usual process of rationalisation and write down the G.S.

****

39A

$$P.I. = \text{Re}\{\dfrac{2(-7 - 9i)}{(-7)^2 - (9i)^2} (\cos 3x + i \sin 3x)\}$$

$$= \dfrac{1}{65}(9 \sin 3x - 7 \cos 3x)$$

therefore G.S. is $y = Ae^{-x} + Be^{-2x} + \dfrac{1}{65}(9 \sin 3x - 7 \cos 3x)$

as obtained before in FRAME 37A.

SECOND ORDER DIFFERENTIAL EQUATIONS

FRAME 40

For practice in using this alternative method, solve the d.e. (38.1)
$$y'' - 2y' + y = 5 \sin 2x.$$

****

40A

C.F. is $(A + Bx)e^x$ *as before.*
$Q = 5 \sin 2x = Im(5e^{2ix})$, *therefore try* $y = \alpha e^{2ix}$ *for P.I.*
Substitution gives $\alpha = -\dfrac{5}{3 + 4i}$.
Therefore P.I. $= Im\{\dfrac{-5}{3 + 4i} e^{2ix}\}$
$\qquad = \dfrac{1}{5}(4 \cos 2x - 3 \sin 2x)$ *on rationalising.*
G.S. is $y = (A + Bx)e^x + \dfrac{1}{5}(4 \cos 2x - 3 \sin 2x)$ *as in FRAME 38A.*

FRAME 41

The main advantages of this alternative complex exponential method over the trigonometric method are

(a) only one parameter has to be determined instead of two
(b) the differentiation and substitution are more compact.

The complex exponential method becomes even more powerful when considering more complicated forms of Q, such as those arising in the next case.

FRAME 42

Case (iv) $Q = ke^{\lambda x} \cos \mu x$ or $ke^{\lambda x} \sin \mu x$

In a similar manner to FRAME 39 we consider $Q = ke^{\lambda x} \cdot e^{i\mu x} = ke^{(\lambda + i\mu)x}$.
For the P.I. the trial solution $y = \alpha e^{(\lambda + i\mu)x}$ is suggested. α is determined in the usual way and finally Re or Im parts are taken as appropriate.

To demonstrate, we consider the d.e.
$$y'' + y' - 2y = 5e^{-x} \sin 2x. \qquad (42.1)$$

SECOND ORDER DIFFERENTIAL EQUATIONS

FRAME 42 continued

The A.E. is $m^2 + m - 2 = 0$ with roots 1 and -2 and C.F. $Ae^x + Be^{-2x}$.

$$Q = 5e^{-x}\sin 2x = \text{Im}(5e^{-x}.e^{2ix}) = \text{Im}\{5e^{(-1+2i)x}\}$$

$\therefore$ try $y = \alpha e^{(-1+2i)x}$ for P.I. of $y'' + y' - 2y = 5e^{(-1+2i)x}$.

Substituting in the d.e.

$$\alpha e^{(-1+2i)x}\{-2 + (-1 + 2i) + (-1 + 2i)^2\} \equiv 5e^{(-1+2i)x}$$

$\therefore \alpha\{-2 + (-1 + 2i) + (1 - 4i - 4)\} = 5$

giving

$$\alpha = \frac{5}{-6 - 2i} = -\frac{5}{2(3 + i)}$$

$\therefore$ P.I. $= \text{Im}\{-\dfrac{5}{2(3 + i)} e^{(-1+2i)x}\}$

$\qquad\qquad = \text{Im}\{-\dfrac{5(3 - i)}{2.10} e^{-x}(\cos 2x + i \sin 2x)\}$

on rationalising and separating the exponential.

This gives

$$\text{P.I.} = \tfrac{1}{4}e^{-x}(\cos 2x - 3 \sin 2x).$$

Therefore G.S. is $y = Ae^x + Be^{-2x} + \tfrac{1}{4}e^{-x}(\cos 2x - 3 \sin 2x).$

FRAME 43

Now solve the d.e.

$$y'' - 2y' + 5y = e^{2x}\cos x.$$

****

43A

The roots of the A.E. are $1 \pm 2i$ and the C.F. is $e^x(A \cos 2x + B \sin 2x)$.
$Q = e^{2x}\cos x = Re(e^{2x}.e^{ix}) = Re\{e^{(2+i)x}\}.$
Therefore try $y = \alpha e^{(2+i)x}$ for P.I.

SECOND ORDER DIFFERENTIAL EQUATIONS

<u>43A</u> continued

Substitution gives

$$\alpha e^{(2+i)x}\{5 - 2(2 + i) + (2 + i)^2\} \equiv e^{(2+i)x}$$

yielding

$$\alpha = \frac{1}{4 + 2i}$$

therefore P.I. $= Re\{\frac{1}{2(2 + i)} e^{(2+i)x}\}$

whence P.I. $= \frac{1}{10} e^{2x}(2 \cos x + \sin x)$

and G.S. is $y = e^x(A \cos 2x + B \sin 2x) + \frac{1}{10} e^{2x}(2 \cos x + \sin x).$

<u>FRAME 44</u>

We have suggested the use of the complex exponential method in Case (iv) where $Q = ke^{\lambda x}\cos \mu x$. It is particularly convenient since the trial solution reduces to a <u>single</u> exponential term. The trigonometric method could, of course, be used in this case but it would be found to be tedious, the reason being that the trial solution would have to be of the form

$$y = e^{\lambda x}(\alpha \cos \mu x + \beta \sin \mu x)$$

since both sine and cosine terms arise on differentiation. <u>Two</u> parameters α and β have to be found and furthermore double differentiation of the products is lengthy. However, for comparison, we will solve the d.e. (42.1)

$$y'' + y' - 2y = 5e^{-x}\sin 2x$$

using this method.

The C.F. is $(Ae^x + Be^{-2x})$ as before.

For the P.I. try $y = e^{-x}(\alpha \sin 2x + \beta \cos 2x).$

FRAME 44 continued

Substituting

$$-2\{e^{-x}(\alpha \sin 2x + \beta \cos 2x)\}$$

$$+ \{e^{-x}(2\alpha \cos 2x - 2\beta \sin 2x) - e^{-x}(\alpha \sin 2x + \beta \cos 2x)\}$$

$$+ \{e^{-x}(-4\alpha \sin 2x - 4\beta \cos 2x) - 2e^{-x}(2\alpha \cos 2x - 2\beta \sin 2x)$$

$$+ e^{-x}(\alpha \sin 2x + \beta \cos 2x)\}$$

$$\equiv 5e^{-x}\sin 2x.$$

Now equate the coefficients of sin 2x and cos 2x, solve for α and β to obtain the P.I., and complete the solution.

****

44A

Coefficient of sin 2x $-2\alpha - 2\beta - \alpha - 4\alpha + 4\beta + \alpha = 5$

or $-6\alpha + 2\beta = 5.$

Coefficient of cos 2x $-2\beta + 2\alpha - \beta - 4\beta - 4\alpha + \beta = 0$

or $-2\alpha - 6\beta = 0.$

Solving these we find $\alpha = -\frac{3}{4}$ *and* $\beta = \frac{1}{4}.$

P.I. is $\frac{1}{4} e^{-x}(\cos 2x - 3 \sin 2x).$

G.S. is $y = Ae^{x} + Be^{-2x} + \frac{1}{4} e^{-x}(\cos 2x - 3 \sin 2x)$ *as in FRAME 42A.*

You will now appreciate that the complex exponential method has the advantages described earlier and, hence, the trigonometric method is <u>not recommended</u>.

FRAME 45

Sum and product forms of Q

Q may sometimes be a combination of two (or more) of the types of functions discussed in Cases (i) - (iv). We shall illustrate by considering the following examples:

SECOND ORDER DIFFERENTIAL EQUATIONS

FRAME 45 continued

$$y'' + 2y' + y = x + \cos x \qquad (45.1)$$
$$y'' + 2y' + y = x \cos x \qquad (45.2)$$

In (45.1) Q is the sum of two functions, i.e. $Q = Q_1 + Q_2$, and in (45.2) Q is the product of two functions, i.e. $Q = Q_1 \cdot Q_2$, where $Q_1 = x$ and $Q_2 = \cos x$.

If ϕ_1 is the P.I. corresponding to Q_1 alone on the R.H.S. of the d.e. and ϕ_2 is that corresponding to Q_2 alone, then by an argument similar to that used in FRAME 16A, it can be shown that the complete P.I. ϕ corresponding to $Q = Q_1 + Q_2$ is $\phi = \phi_1 + \phi_2$. This is <u>not</u> true in the case of a product. These statements will now be further clarified by solving the equations given in this frame.

FRAME 46

Starting with (45.1)
$$y'' + 2y' + y = x + \cos x,$$
it is seen that the C.F. is $(A + Bx)e^{-x}$.

For the P.I. for $Q_1 = x$, try $y = \alpha x + \beta$,

and for the P.I. for $Q_2 = \cos x = \text{Re}(e^{ix})$, try $y = \gamma e^{ix}$.

Determine the values of α, β and γ by the usual methods.

****

46A

For ϕ_1, substituting $y = \alpha x + \beta$, we have
$$(\alpha x + \beta) + 2.\alpha \equiv x$$
giving $\alpha = 1$ and $\beta = -2$ on equating the coefficients.
Therefore P.I. for x is $\phi_1 = x - 2$.
For ϕ_2, substituting $y = \gamma e^{ix}$,
$$\gamma e^{ix}\{1 + 2i + (i)^2\} \equiv e^{ix}$$
$$\therefore \gamma = \frac{1}{2i}$$

<u>46A continued</u>

$\therefore$ P.I. for $\cos x$ is $\phi_2 = Re\left(\frac{1}{2i} e^{ix}\right) = \frac{1}{2} \sin x$.

$\therefore$ P.I. for $x + \cos x$ is $\phi = \phi_1 + \phi_2 = (x - 2) + \frac{1}{2} \sin x$, *giving for the G.S.* $y = (A + Bx)e^{-x} + x - 2 + \frac{1}{2} \sin x$.

<u>FRAME 47</u>

Now consider (45.2)
$$y'' + 2y' + y = x \cos x,$$
the C.F. of which is again $(A + Bx)e^{-x}$.

In this case the method of obtaining ϕ_1 and ϕ_2 separately and then taking the product is <u>not</u> valid. We require instead the complete P.I. for
$Q = x \cos x = Re(xe^{ix})$.

We therefore try $y = (\alpha x + \beta)e^{ix}$.
Substitution yields
$$e^{ix}(\alpha x + \beta) + 2e^{ix}\{\alpha + i(\alpha x + \beta)\} + e^{ix}\{i\alpha + i\alpha - (\alpha x + \beta)\} \equiv xe^{ix}$$
or $\qquad 2i(\alpha x + \beta) + (2\alpha + 2i\alpha) \equiv x.$

α and β are determined by equating the coefficients
giving $\qquad \alpha = \frac{1}{2i} = -\frac{i}{2}$ and $\beta = \frac{i}{2} + \frac{1}{2}$.

$\therefore$ P.I. for $x \cos x$ is $\phi = Re\{\frac{1}{2}(-ix + i + 1)(e^{ix})\}$.

Simplify and write down the G.S.

****

<u>47A</u>

G.S. is $y = (A + Bx)e^{-x} + \frac{1}{2}\{\cos x + (x - 1)\sin x\}$.

Note that this P.I. is certainly <u>not</u> *the product of the two separate P.I.'s given in FRAME 46.*

SECOND ORDER DIFFERENTIAL EQUATIONS

FRAME 48

For extra practice, solve the following d.e.'s:

(1) $\quad y'' + 8y' + 17y = 2e^{-3x}$

(2) $\quad y'' + 4y' + 4y = 8x^2 - 5\cos x$

(3) $\quad 2y'' + y' - y = 3e^{-x}\sin 2x$

(4) $\quad y'' - 2y' = 5xe^{-x}$

****

48A

(1) $\quad y = e^{-4x}(A\cos x + B\sin x) + e^{-3x}$

(2) $\quad y = (A + Bx)e^{-2x} + (2x^2 - 4x + 3) - \frac{1}{5}(3\cos x + 4\sin x)$

(3) $\quad y = Ae^{\frac{1}{2}x} + Be^{-x} + \frac{3}{50}e^{-x}(3\cos 2x - 4\sin 2x)$

(4) $\quad y = A + Be^{2x} + \frac{5}{9}(3x + 4)e^{-x}$

FRAME 49

Cases of failure

In some cases, due to the structure of the d.e., some of the functional forms arising in the trial solution appear in the C.F. For instance, consider the d.e.
$$y'' - y' - 2y = 5e^{2x}. \qquad (49.1)$$

The C.F. is $Ae^{2x} + Be^{-x}$ which contains the function e^{2x} appearing in Q.

At first sight, $y = \alpha e^{2x}$ is suggested as a trial solution to obtain the P.I. However, on substitution we obtain
$$\alpha e^{2x}(-2 - 2 + 4) \equiv 5e^{2x}$$

showing that α is indeterminate. This is due to the fact that e^{2x} is a solution of the reduced equation. Such cases are referred to as <u>cases of failure</u>. In physical problems they correspond to the phenomenon of resonance

FRAME 49 continued

as is illustrated by Example 11 of FRAME 67. It is easier to solve these equations using either D-operator or Laplace Transform methods which are covered by later programmes in this series. If you are going to read these programmes, we suggest you proceed to FRAME 55. If not, continue with the next frame.

FRAME 50

You will recall from FRAMES 20 - 22 that when this repetition of functions occurred the remedy was to multiply the recurring one by x. A similar device is used in cases of failure. So the trial solution for the P.I. in (49.1) is modified to $y = \alpha x e^{2x}$, on multiplying by x. (Note that this case of failure could only have been anticipated by finding the C.F. first.)

Now substituting the modified trial solution we have
$$-2\alpha x e^{2x} - \alpha e^{2x}(1 + 2x) + \alpha e^{2x}(2 + 2 + 4x) \equiv 5e^{2x}.$$

The terms in xe^{2x} vanish (as they must do if this trial solution is to be successful) giving
$$3\alpha e^{2x} \equiv 5e^{2x}$$
and hence
$$\alpha = \frac{5}{3}.$$

The P.I. is $\frac{5}{3} x e^{2x}$ and the G.S. is $y = (Ae^{2x} + Be^{-x}) + \frac{5}{3} x e^{2x}$.

FRAME 51

You have seen in the previous frame how to deal with the case of failure when Q is exponential. We shall now discuss the following examples in which the three remaining forms of Q (listed in FRAME 32) occur involving cases of failure:

$$y'' + 4y' = 2x \qquad (51.1)$$
$$y'' + 4y = 3 \sin 2x \qquad (51.2)$$
$$y'' - 4y' + 5y = 2e^{2x} \sin x \qquad (51.3)$$

SECOND ORDER DIFFERENTIAL EQUATIONS

FRAME 52

In (51.1), $y'' + 4y' = 2x$,
the A.E. is $m^2 + 4m = 0$ with roots $m = 0$ and -4 giving the C.F. $A + Be^{-4x}$.

Normally the trial solution for the P.I. would be $y = \alpha x + \beta$. However the constant term already appears in the C.F. Hence there is a case of failure suggesting the modified trial solution

$$y = x.(\alpha x + \beta) = \alpha x^2 + \beta x.$$

Substitute, determine the values of α and β by equating the coefficients, and complete the solution.

****

52A

$$4(2\alpha x + \beta) + 2\alpha \equiv 2x$$

from which $\alpha = \frac{1}{4}$ *and* $\beta = -\frac{1}{8}$

$\therefore$ *P.I. is* $\frac{1}{4}x^2 - \frac{1}{8}x.$

$\therefore$ *G.S. is* $y = A + Be^{-4x} + \frac{x}{8}(2x - 1).$

FRAME 53

Now consider (51.2), $y'' + 4y = 3 \sin 2x$.

The A.E. is $m^2 + 4 = 0$ with roots $\pm 2i$ and C.F. $A \cos 2x + B \sin 2x$. The presence of $\sin 2x$ in the C.F. indicates a case of failure.

Therefore for the P.I.: $3 \sin 2x = \text{Im}(3e^{2ix})$, and the trial solution $y = \alpha e^{2ix}$ has to be modified to
$$y = \alpha x e^{2ix}.$$

Complete the solution.

****

53A

Substitution gives

$$\alpha e^{2ix}(4i - 4x) + 4\alpha x e^{2ix} \equiv 3e^{2ix}$$

giving $\alpha = \dfrac{3}{4i}$.

$\therefore$ P.I. $= \text{Im}(\dfrac{3}{4i} x e^{2ix}) = -\dfrac{3}{4}x \cos 2x$

$\therefore$ G.S. is $y = A \cos 2x + B \sin 2x - \dfrac{3}{4}x \cos 2x$.

FRAME 54

Finally, in (51.3), $y'' - 4y' + 5y = 2e^{2x}\cos x$, the A.E. is $m^2 - 4m + 5 = 0$ with roots $2 \pm i$, and C.F. $e^{2x}(A \cos x + B \sin x)$.

The $e^{2x}\cos x$ term in the C.F. indicates a case of failure.

$\therefore$ for the P.I.: $2e^{2x}\cos x = \text{Re}\{2e^{(2+i)x}\}$, and the trial solution $y = \alpha e^{(2+i)x}$ has to be modified to

$$y = \alpha x e^{(2+i)x}.$$

Now complete the solution as before.

****

54A

$$5\alpha x e^{(2+i)x} - 4\alpha e^{(2+i)x}\{1 + (2 + i)x\}$$

$$+ \alpha e^{(2+i)x}\{(2 + i) + (2 + i) + (2 + i)^2 x\} \equiv 2e^{(2+i)x}$$

The x terms cancel (as they must) yielding

$2i\alpha = 2$ or $\alpha = 1/i$.

$\therefore$ P.I. $= \text{Re}\{\dfrac{1}{i} x e^{(2+i)x}\} = x e^{2x}\sin x$

$\therefore$ G.S. is $y = e^{2x}(A \cos x + B \sin x) + x e^{2x}\sin x$.

SECOND ORDER DIFFERENTIAL EQUATIONS

FRAME 55

Boundary Conditions

In physical problems, solutions are usually required which satisfy certain specified conditions. These conditions provide information from which values may be assigned to the arbitrary constants. This type of solution which satisfies certain definite conditions is called a PARTICULAR SOLUTION and the conditions satisfied are called BOUNDARY CONDITIONS or INITIAL CONDITIONS.

In the type of d.e. considered in this programme, two arbitrary constants appear in the general solution. Naturally, two boundary or initial conditions will be required.

FRAME 56

For example, let us consider the d.e. (42.1)
$$y'' + y' - 2y = 5e^{-x}\sin 2x.$$
We obtained the G.S. as
$$y = Ae^x + Be^{-2x} + \tfrac{1}{4}e^{-x}(\cos 2x - 3 \sin 2x)$$
where A and B are the arbitrary constants. If we are given the two boundary conditions $y = 1$ and $y' = 0$ when $x = 0$, A and B can be determined as follows.
$$y = 1 \quad \text{when} \quad x = 0 \quad \text{gives}$$
$$1 = A + B + \tfrac{1}{4}.$$
On differentiating the G.S.
$$y' = Ae^x - 2Be^{-2x} + \tfrac{1}{4}e^{-x}(-2 \sin 2x - 6 \cos 2x) - \tfrac{1}{4}e^{-x}(\cos 2x - 3 \sin 2x)$$
and applying $y' = 0$ when $x = 0$ gives
$$0 = A - 2B - \tfrac{6}{4} - \tfrac{1}{4}.$$
We therefore have the two simultaneous equations
$$A + B = \tfrac{3}{4}$$
$$A - 2B = \tfrac{7}{4}$$

D

FRAME 56 continued

$$\text{from which} \quad A = \frac{13}{12} \quad \text{and} \quad B = -\frac{1}{3}.$$

Therefore the required Particular Solution is

$$y = \frac{13}{12} e^x - \frac{1}{3} e^{-2x} + \frac{1}{4} e^{-x}(\cos 2x - 3 \sin 2x).$$

Note that the boundary conditions must be applied to the complete G.S. and <u>not</u> to the C.F. part only.

FRAME 57

Find the particular solutions to the d.e.'s (1) and (3) of FRAME 48 subject to the boundary conditions:

$$y = 2 \quad \text{when} \quad x = 0 \quad \text{and} \quad y = 0 \quad \text{when} \quad x = \frac{\pi}{2} \quad \text{for (1)}$$

and

$$y = 1 \quad \text{and} \quad y' = 0 \quad \text{when} \quad x = 0 \quad \text{for (3)}.$$

****

57A

(1) $\quad y = e^{-4x}(\cos x - e^{\pi/2} \sin x) + e^{-3x}$

(3) $\quad y = \frac{74}{75} e^{\frac{1}{2}x} - \frac{1}{6} e^{-x} + \frac{3}{50} e^{-x}(3 \cos 2x - 4 \sin 2x).$

FRAME 58

Applications

We shall now consider some examples in which the type of equation studied in this programme is applied to physical problems.

For instance, in an L,R,C series circuit an inductance L H, a resistance R Ω and a condenser of capacitance C F are connected in series to a generator supplying an e.m.f. E(t) V which may vary with time t.

SECOND ORDER DIFFERENTIAL EQUATIONS   2:35

FRAME 58 continued

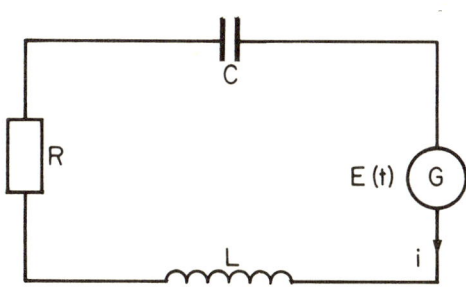

If i A is the current flowing in the circuit and q C is the charge on the condenser plates at time t, then on equating the total potential drop across each of the components to E(t) we have the d.e.

$$L \frac{di}{dt} + Ri + \frac{q}{C} = E(t).$$

Noting that $i = \frac{dq}{dt}$ this becomes

$$L \frac{d^2q}{dt^2} + R \frac{dq}{dt} + \frac{q}{C} = E(t) \qquad (58.1)$$

which is a d.e. for the charge q of precisely the form given in (6.1), L, R and C being assumed independent of t.

Two particular cases of this circuit equation will now be discussed as examples.

FRAME 59

Case (i)

A battery of constant voltage E_o is connected to an inductance L in series with a capacitance C. Show that when the current reverses (next becomes zero) the condenser is charged to a voltage $2E_o$.

In this case, R = 0 and $E(t) = E_o$ (constant).
The equation becomes

$$\frac{d^2q}{dt^2} + \omega^2 q = \frac{E_o}{L} \qquad \left(\omega^2 = \frac{1}{LC}\right)$$

The A.E. is $m^2 + \omega^2 = 0$ with roots $\pm j\omega$ and C.F. $A \cos \omega t + B \sin \omega t$.
(j is used here for $\sqrt{-1}$ to avoid confusion with current i.)

For the P.I. for $\frac{E_o}{L}$ we try $q = \alpha$. Substitution gives

$$\omega^2 \alpha = \frac{E_o}{L} \quad \text{or} \quad \alpha = E_o C.$$

FRAME 59 continued

The general solution is therefore

$$q = A \cos \omega t + B \sin \omega t + E_o C.$$

Now apply the condition that initially (i.e. when t = 0) q and i are both zero and find the particular solution for q and also, by differentiation, for i.

****

59A

$$q = 0 \quad at \quad t = 0 \quad gives \quad 0 = A + E_o C$$

$$i.e. \quad A = - E_o C.$$

Since $i = \dfrac{dq}{dt} = - \omega A \sin \omega t + \omega B \cos \omega t,$

$$i = 0 \quad at \quad t = 0 \quad gives \quad B = 0.$$

Whence charge $q = E_o C(1 - \cos \omega t)$ and current $i = \omega E_o C \sin \omega t.$

The current is next zero (after t = 0) when $\omega t = \pi$

and at this time the charge $q = 2 E_o C.$

$\therefore$ Condenser is charged to voltage $\dfrac{q}{C} = 2 E_o.$

FRAME 60

Case (ii)

A series circuit consists of an inductance of 5 mH (0·005 H), a resistance of 10 Ω and a condenser of capacitance 100 μF (10^{-4} F). The applied e.m.f. is 50 sin 1000t V. Find the charge q on the condenser and the current i flowing in the circuit at time t. Find also the maximum value of the current when the steady state has been reached (i.e. as $t \to \infty$).

Here, L = 0·005, R = 10, C = 10^{-4} and E(t) = 50 sin 1000t giving the d.e.

$$\frac{d^2q}{dt^2} + 2 \times 10^3 \frac{dq}{dt} + 2 \times 10^6 q = 10^4 \sin 1000t$$

on dividing by 0·005.

SECOND ORDER DIFFERENTIAL EQUATIONS

FRAME 60 continued

The roots of the A.E. are $-10^3 \pm 10^3 j$, so that the C.F. is $e^{-1000t}(A \cos 1000t + B \sin 1000t)$.

This is an oscillation of frequency $\dfrac{1000}{2\pi}$ equal to that of the applied e.m.f. and with amplitude heavily damped with the e^{-1000t} factor.

For the P.I. for $10^4 \sin 1000t = \text{Im}(10^4 e^{1000jt})$ we try $q = \alpha e^{1000jt}$ in the usual way.

Find α and complete the general solution for q. Then apply the initial conditions and obtain expressions for q and i at time t.

****

60A

Substituting

$$\alpha e^{1000jt}\{2 \times 10^6 + 2 \times 10^3 \cdot 1000j + (1000j)^2\} \equiv 10^4 e^{1000jt}$$

from which $\alpha = \dfrac{10^{-2}}{1 + 2j}$

$\therefore$ P.I. $= \text{Im}\left\{\dfrac{10^{-2}}{1 + 2j} e^{1000jt}\right\}$

$\qquad\qquad = 2 \times 10^{-3}(\sin 1000t - 2 \cos 1000t).$

The G.S. is therefore

$q = e^{-1000t}(A \cos 1000t + B \sin 1000t) + 2 \times 10^{-3}(\sin 1000t - 2 \cos 1000t)$

from which

$i = \dfrac{dq}{dt} = e^{-1000t}\{-1000A \sin 1000t + 1000B \cos 1000t - 1000A \cos 1000t - 1000B \sin 1000t\} + 2(\cos 1000t + 2 \sin 1000t).$

Applying the boundary conditions $q = 0$ *and* $i = 0$ *at* $t = 0$ *gives* $A = 4 \times 10^{-3}$ *and* $B = 2 \times 10^{-3}$.

$\therefore\ q = 10^{-3}\{e^{-\omega t}(4 \cos \omega t + 2 \sin \omega t) + 2 \sin \omega t - 4 \cos \omega t\}$

and $i = e^{-\omega t}(-2 \cos \omega t - 6 \sin \omega t) + 2 \cos \omega t + 4 \sin \omega t$

where $\omega = 1000$ *for convenience.*

FRAME 61

The "transient" terms represented by the C.F. contribution rapidly disappear as t increases, due to the e^{-1000t} term, and we then obtain the "steady-state" current (given by the P.I.)

$$i = 2 \cos \omega t + 4 \sin \omega t.$$

Using the method of FRAME 26 this may be written more conveniently as

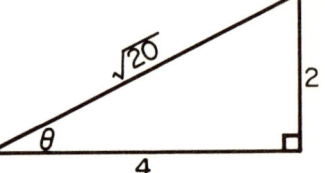

$$i = \sqrt{20} \sin (\omega t + \theta).$$

This shows that the current is in advance of the applied e.m.f. by the phase angle $\theta = \tan^{-1} \frac{1}{2} = 26°34'$ and has amplitude $= \sqrt{20}$. This is, in fact, the maximum value of i,

i.e. maximum current $= \sqrt{20} = 4 \cdot 47$ A.

FRAME 62

As an application to the problem of mechanical vibrations we consider the system in which a 1 kg mass hangs at rest on a spring whose stiffness is such that a force of ·024 N extends it 1 mm. The upper end of the spring is now given a vibrating displacement y = sin nt, y being measured vertically downwards in metres. If the mass is subject to a frictional resistance whose magnitude in newtons is equal to four times the velocity in m/s, find the amplitude of the steady-state oscillation as a function of n. Hence show that the ratio of resonant frequency (for which the amplitude is a maximum) to the natural frequency is approximately 89%. (NOTE: Simple arithmetical calculations, rather than realism, have influenced our choice of numerical values in this problem.)

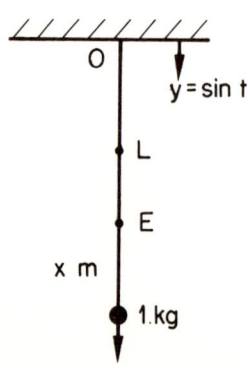

OL is the natural length of the spring, E is the equilibrium position, and x is the displacement from E, at time t s. The effective displacement is (x − sin nt) m and the restoring force is s(x − sin nt) N, where s,

SECOND ORDER DIFFERENTIAL EQUATIONS

FRAME 62 continued

the stiffness of the spring, is 24 N/m. Hence, the d.e. of motion is
$\ddot{x} = -4\dot{x} - 24(x - \sin nt)$, the dots representing differentiations w.r.t. t.

i.e. $\quad \ddot{x} + 4\dot{x} + 24x = 24 \sin nt.$ \hfill (62.1)

The A.E. $m^2 + 4m + 24 = 0$ has roots $-2 \pm 2\sqrt{5}i$ and the C.F. is therefore

$$e^{-2t}(A \cos 2\sqrt{5}t + B \sin 2\sqrt{5}t). \quad (62.2)$$

This is a damped oscillation of frequency $\frac{2\sqrt{5}}{2\pi}$ i.e. $\frac{\sqrt{5}}{\pi}$ c/s, representing the natural frequency of the damped oscillations of the system. (In the absence of damping the natural frequency would be $\frac{\sqrt{24}}{2\pi}$ c/s, as can be seen by solving the reduced equation $\ddot{x} + 24x = 0$.)

The C.F. decays rapidly because of the e^{-2t} factor, providing the transient disturbance, and the steady-state oscillation due to the presence of the "forcing term" $\sin nt$ is given by the P.I.

For the P.I. for $24 \sin nt = \text{Im}(24e^{int})$, try $y = \alpha e^{int}$. Find the parameter α and write the steady-state solution in the form

$$R \sin(nt - \theta).$$

****

62A

Substitution gives
$$\alpha e^{int}[24 + 4in - n^2] = 24e^{int}$$

from which
$$\alpha = \frac{24}{(24 - n^2) + 4in}$$

Therefore P.I. $= \text{Im}\left[\frac{24e^{int}}{(24 - n^2) + 4in}\right]$

$$= \frac{24[(24 - n^2)\sin nt - 4n \cos nt]}{(24 - n^2)^2 + (4n)^2}$$

on rationalisation.

The steady-state solution may therefore be written in the form

$$x = R \sin(nt - \theta)$$

2:40   SECOND ORDER DIFFERENTIAL EQUATIONS

62A continued

where, from the diagram

$$\theta = \tan^{-1}\left[4n/(24 - n^2)\right]$$

and the amplitude is given as a function of n by

$$R = \frac{24}{\left[(24 - n^2)^2 + (4n)^2\right]^{\frac{1}{2}}}.$$

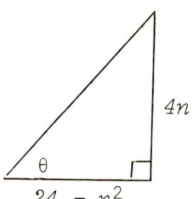

FRAME 63

This solution $x = R \sin(nt - \theta)$, represents a "forced" oscillation of frequency $n/2\pi$ equal to that of the applied vibration but with a phase lag θ. The resonant frequency is obtained by maximising the amplitude R. This implies that

$$\left[(24 - n^2)^2 + (4n)^2\right]$$

is a minimum, therefore

$$\frac{d}{dn^2}\left[(24 - n^2)^2 + 16n^2\right] = 0,$$

differentiating w.r.t. n^2 for convenience.

$$\therefore -2(24 - n^2) + 16 = 0$$

$$\text{i.e. } n^2 = 16$$

$$\therefore \text{Resonant frequency} = \frac{4}{2\pi} \text{ c/s.}$$

Ratio of resonant to natural frequency is

$$\frac{2/\pi}{\sqrt{5}/\pi} = 0 \cdot 894 \simeq 89\%$$

FRAME 64

As a final example we consider the problem of the buckling of a strut. Here, a light uniform beam of length ℓ is clamped horizontally at both ends. It is loaded uniformly with weight w per unit length and is subject to a compressive load P at its ends. Find the magnitude of the clamping couple, G.
(NOTE: Perhaps we should make it clear that, in S.I. units, w would be in newtons per metre, P in newtons, and so on.)

SECOND ORDER DIFFERENTIAL EQUATIONS

FRAME 64 continued

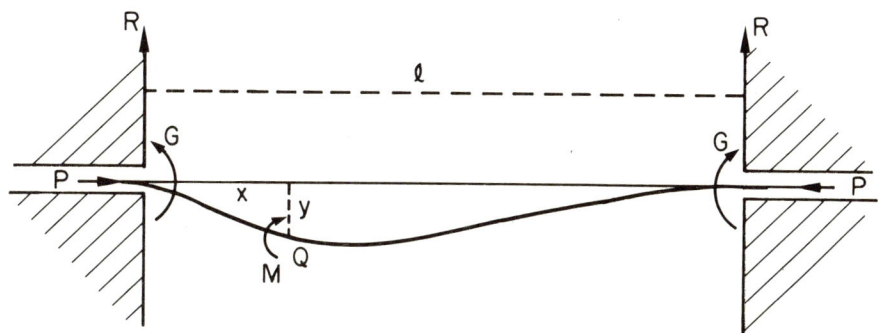

Taking moments about Q for forces to the left of Q we obtain

$$(M - G) + Rx + Py - wx \cdot \frac{x}{2} = 0$$

where $M (= EIy'')$ is the bending moment, the reaction R is equal to $w\ell/2$ and the load wx of the portion of length x has been assumed to act at the mid-point $x/2$. This gives the d.e.

$$EIy'' + Py = G - \tfrac{1}{2}w\ell x + \tfrac{1}{2}wx^2$$

for the deflection y at position x, EI being the constant flexural rigidity of the beam.

Writing this as

$$y'' + n^2 y = \frac{1}{EI}(G - \tfrac{1}{2}w\ell x + \tfrac{1}{2}wx^2),$$

where $n^2 = P/EI$, we see that the C.F. is

$$A \cos nx + B \sin nx.$$

For the particular integral try $y = \alpha x^2 + \beta x + \gamma$.

Find the parameters α, β and γ and write down the general solution. Then find the arbitrary constants A and B by applying the boundary conditions $y = y' = 0$ at the end $x = 0$ which is clamped horizontally.

****

64A

Substitution gives

$$n^2(\alpha x^2 + \beta x + \gamma) + 2\alpha \equiv \frac{1}{EI}(G - \tfrac{1}{2}w\ell x + \tfrac{1}{2}wx^2)$$

and equating the coefficients we obtain

$$\alpha = \frac{w}{2P}, \qquad \beta = -\frac{w\ell}{2P} \qquad \text{and} \qquad \gamma = \frac{1}{P}(G - \frac{w}{n^2})$$

on noting that $\dfrac{1}{EI} = \dfrac{n^2}{P}$.

Therefore the general solution is

$$y = A\cos nx + B\sin nx + \frac{1}{P}(G - \frac{w}{n^2} - \tfrac{1}{2}w\ell x + \tfrac{1}{2}wx^2).$$

Applying the boundary conditions $y = y' = 0$ *at* $x = 0$ *gives*

$$A = \frac{1}{P}(\frac{w}{n^2} - G)$$

$$\text{and} \quad B = \frac{w\ell}{2Pn}$$

yielding

$$y = \frac{1}{P}\{(\frac{w}{n^2} - G)\cos nx + \frac{w\ell}{2n}\sin nx + G - \frac{w}{n^2} - \tfrac{1}{2}w\ell x + \tfrac{1}{2}wx^2\}.$$

FRAME 65

Now find the value of the clamping couple G by applying the hitherto unused boundary condition $y = 0$ at the other end, i.e. where $x = \ell$.

****

65A

Substituting the condition $y = 0$ *when* $x = \ell$ *in the above particular solution we find*

$$G - \frac{w}{n^2} = -\left(\frac{w\ell}{2n}\sin n\ell\right)/(1 - \cos n\ell)$$

and simplification produces

$$G = \frac{w}{n^2}\left(1 - \frac{n\ell}{2}\cot\frac{n\ell}{2}\right).$$

You may care to verify that the remaining boundary condition $y' = 0$ *at* $x = \ell$ *is automatically satisfied now that G has been found.*

SECOND ORDER DIFFERENTIAL EQUATIONS

FRAME 66

Summary

Before trying some miscellaneous examples you may find it useful to refer to the following summary of the various methods of obtaining C.F.'s and P.I.'s. Omit the section on cases of failure if you did not read FRAMES 50 - 54.

The Complementary Function

Case (i)   A.E. has real unequal roots m_1 and m_2
           C.F. is $Ae^{m_1 x} + Be^{m_2 x}$.

Case (ii) A.E. has real equal roots $m_1 = m_2 = n$
           C.F. is $(A + Bx)e^{nx}$.

Case (iii) A.E. has conjugate complex roots $p \pm iq$
           C.F. is $e^{px}(A \cos qx + B \sin qx)$.

The Particular Integral

The principle used in obtaining a trial solution for the P.I. for the right-hand side, Q, is to include a multiple of Q together with any functions which may arise on differentiating twice. The main cases are tabulated as follows:

Form of Q	Suggested Trial Solution
$ke^{\lambda x}$	$y = \alpha e^{\lambda x}$
$\lambda x^2 + \mu x + \nu$	$y = \alpha x^2 + \beta x + \gamma$
$k \cos \mu x$ $k \sin \mu x$	$y = \alpha e^{i\mu x}$ and take Re or Im parts
$ke^{\lambda x} \cos \mu x$ $ke^{\lambda x} \sin \mu x$	$y = \alpha e^{(\lambda + i\mu)x}$ and take Re or Im parts.

Cases of failure

When a term of the indicated trial solution already appears in the C.F., modify the trial solution by multiplying by x <u>before</u> substitution.

SECOND ORDER DIFFERENTIAL EQUATIONS

FRAME 67

Miscellaneous Examples

In this frame a collection of miscellaneous examples is given for you to try. Answers are supplied in FRAME 68 and hints have been provided in some cases.

Solve the following d.e.'s, subject to the conditions given where appropriate. Some practical applications have been included.

If you omitted FRAMES 50 - 54 you will be able to identify cases of failure, but not solve them.

1. $y'' + 4y' + 5y = 3e^{-2x}$ given that $y = 4$ and $y' = -7$ at $x = 0$

2. $4y'' - 4y' + y = x^2 - 3x - 13$

3. $y'' + 4y' + 4y = 2e^x - 3\cos x$

4. $y'' - y = x \sin x$

5. $y'' - 3y' + 2y = \frac{1}{2}e^x$ given that $y = 0$ and $y' = 0$ when $x = 0$

6. $y'' + y' = x$ given that $y = y' = 1$ when $x = 0$

7. $y'' + 9y = 6\cos 3x$ subject to $y = \frac{\pi}{4}$ when $x = 0$ and $y = \frac{\pi}{3}$ when $x = \frac{\pi}{6}$

8. $y'' - 2y' + 5y = e^{-x}\sin 2x + e^x \sin 2x$

9. The equation of motion for a simple pendulum of length ℓ is

$$\frac{d^2\theta}{dt^2} = -\frac{g}{\ell}\theta$$

where θ is the angular displacement (assumed small) at time t. If the system is released from rest in the position $\theta = \theta_o$ find the subsequent displacement and show that the periodic time is

$$T = 2\pi\sqrt{\frac{\ell}{g}}.$$

SECOND ORDER DIFFERENTIAL EQUATIONS 2:45

FRAME 67 continued

10. 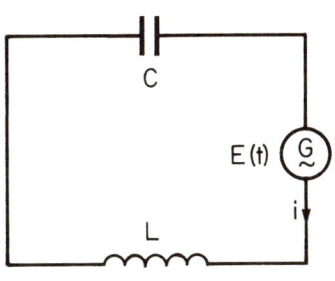 In an oscillator circuit the d.e. for the charge q on the condenser at time t is

$$L\ddot{q} + \frac{q}{C} = E(t). \qquad \text{(cf. FRAME 58)}$$

If the applied e.m.f. $E(t)$ has the form $E_o \cos pt$, find q and the current i at time t in the case $p \neq \omega$ where $\omega^2 = 1/LC$, if both q and i are zero initially.

11. If the circuit of Example 10 is tuned so that $p = \omega$ show that the expressions for q and i are modified to

$$q = \frac{E_o}{2\omega L} t \sin \omega t$$

and $$i = \frac{E_o}{2\omega L} (\sin \omega t + \omega t \cos \omega t).$$

(Note that because of the presence of the t factor the amplitudes become very large. This is the physical phenomenon of "resonance" which occurs when we have a tuned circuit, and corresponds to the mathematical "case of failure".)

12. A body of unit mass moves along the x-axis under the action of an attractive force of magnitude $\omega^2 x$ directed towards the origin O. It is subject also to a frictional resistance of magnitude $2kv$, v being the velocity at time t, and to a driving force $e^{-kt} \cos pt$. The equation of motion, may, therefore, be written as

$$\ddot{x} + 2k\dot{x} + \omega^2 x = e^{-kt} \cos pt.$$

If the resistance is sufficiently small, so that $k < \omega$, we may write $\omega^2 - k^2 = n^2$, where n is real. Obtain the solution of the equation when

(i) $p \neq n$

and (ii) $p = n$,

assuming in both cases that the body starts from rest at the origin.

FRAME 67 continued

13. A uniform beam of length ℓ and weight w per unit length is freely supported horizontally at its end. It bends under the action of its own weight and a horizontal compressive thrust P, applied at the ends. Calculate the sag at the centre of the beam, assuming that the displacement y at position x is given by

$$EIy'' + Py = \tfrac{1}{2}wx^2 - \tfrac{1}{2}w\ell x. \qquad \text{(cf. FRAME 64)}$$

FRAME 68

Answers to Miscellaneous Examples

1. $y = e^{-2x}(\cos x + \sin x) + 3e^{-2x}$

2. $y = (A + Bx)e^{\frac{1}{2}x} + (x^2 + 5x - 1)$

3. $y = (A + Bx)e^{-2x} + \tfrac{2}{9}e^x - \tfrac{3}{25}(3\cos x + 4\sin x)$

4. $y = Ae^x + Be^{-x} - \tfrac{1}{2}(\cos x + x \sin x)$

   $\{$Try $y = (\alpha x + \beta)e^{ix}$ and take Im parts for P.I.$\}$

5. Case of failure, $y = \tfrac{1}{2}(e^{2x} - e^x - xe^x)$

6. Case of failure. $y = 3 - 2e^{-x} + \tfrac{1}{2}x^2 - x$

7. Case of failure. $y = \tfrac{\pi}{4}\cos 3x + \tfrac{\pi}{6}\sin 3x + x \sin 3x$

   $\{$Try $y = \alpha x e^{3ix}$ and take Re parts for P.I.$\}$

8. $y = e^x(A\cos 2x + B\sin 2x) + \tfrac{1}{20}e^{-x}(\sin 2x + 2\cos 2x) - \tfrac{1}{4}xe^x \cos 2x$

   $\{$For first P.I. try $y = \alpha e^{(-1+2i)x}$ and take Im parts

   $\alpha = \dfrac{1}{4(1-2i)}$

   For second P.I. try $y = \beta x e^{(1+2i)x}$ (case of failure) and take Im parts

   $\beta = \dfrac{1}{4i}\}$

SECOND ORDER DIFFERENTIAL EQUATIONS

FRAME 68 continued

9. Rewriting the equation as $\ddot{\theta} + \omega^2\theta = 0$ where $\omega^2 = \frac{g}{\ell}$,
$\theta = A \cos \omega t + B \sin \omega t$, subject to $\theta = \theta_o$, $\theta' = 0$ at $t = 0$.
Hence $\theta = \theta_o \cos \omega t$.

Frequency $= \frac{\omega}{2\pi}$ and $T = \frac{2\pi}{\omega} = 2\pi\sqrt{\frac{\ell}{g}}$.

10. $\ddot{q} + \omega^2 q = \frac{E_o}{L} \cos pt$

G.S. is $q = A \cos \omega t + B \sin \omega t + \frac{E_o}{L(\omega^2 - p^2)} \cos pt$ and

$i = \dot{q} = -A\omega \sin \omega t + B\omega \cos \omega t - \frac{E_o p}{L(\omega^2 - p^2)} \sin pt$.

(Try $q = \alpha e^{jpt}$ for P.I. and take Re parts.)

Applying $q = 0 = i$ at $t = 0$ gives $A = -\frac{E_o}{L(\omega^2 - p^2)}$

and $B = 0$ yielding

$q = \frac{E_o}{L(\omega^2 - p^2)} (\cos pt - \cos \omega t)$ $\qquad (\omega \neq p)$

and $\quad i = \frac{E_o}{L(\omega^2 - p^2)} (\omega \sin \omega t - p \sin pt)$.

11. $\ddot{q} + \omega^2 q = \frac{E_o}{L} \cos \omega t$ $\qquad\qquad (\omega = p)$

G.S. is $q = A \cos \omega t + B \sin \omega t + \frac{E_o}{2\omega L} t \sin \omega t$

and $\quad i = \dot{q} = -A\omega \sin \omega t + B\omega \cos \omega t + \frac{E_o}{2\omega L}(\sin \omega t + \omega t \cos \omega t)$.

$\{$Try $y = \beta t e^{j\omega t}$ (case of failure) for P.I. and take Re parts.$\}$

Applying $q = 0 = i$ at $t = 0$ gives $A = B = 0$ and the result follows.

FRAME 68 continued

12. A.E. is $m^2 + 2km + \omega^2 = 0$.

Roots are $-k \pm \sqrt{k^2 - \omega^2} = -k \pm in$.

Therefore C.F. is $e^{-kt}(A \cos nt + B \sin nt)$.

Case (i) $p \neq n$

For P.I. for $e^{-kt} \cos pt$, try $x = \alpha e^{(-k+ip)t}$.

Substitution gives
$$\alpha\{\omega^2 + 2k(-k + ip) + (-k + ip)^2\} = 1$$

therefore $\alpha = \dfrac{1}{n^2 - p^2}$.

On taking Re parts the P.I. is $\dfrac{1}{n^2 - p^2} e^{-kt} \cos pt$

and the G.S. is $x = e^{-kt}\left(A \cos nt + B \sin nt + \dfrac{\cos pt}{n^2 - p^2}\right)$.

Applying $x = 0$ and $\dot{x} = 0$ at $t = 0$ gives $A = -\dfrac{1}{n^2 - p^2}$ and $B = 0$.

Hence $x = \dfrac{e^{-kt}}{n^2 - p^2}(\cos pt - \cos nt)$.

Case (ii) $p = n$

For P.I. for $e^{-kt} \cos nt$, try $y = \beta t e^{(-k+in)t}$ since this is a case of failure.

Substitution yields
$$\omega^2 \beta t e^{(-k+in)t} + 2k\beta e^{(-k+in)t}\{1 + (-k + in)t\}$$
$$+ \beta e^{(-k+in)t}\{(-k + in) + (-k + in) + (-k + in)^2 t\}$$
$$\equiv e^{(-k+in)t}.$$

Therefore $2in\beta = 1$.

P.I. $= \text{Re } \dfrac{1}{2in} t e^{(-k+in)t} = \dfrac{t}{2n} e^{-kt} \sin nt$

$x = e^{-kt}\left(A \cos nt + B \sin nt + \dfrac{t}{2n} \sin nt\right)$.

Applying $x = 0 = \dot{x}$ at $t = 0$ in this case gives $A = B = 0$.

Therefore $x = \dfrac{t}{2n} e^{-kt} \sin nt$.

SECOND ORDER DIFFERENTIAL EQUATIONS

FRAME 68 continued

13. Rewriting the equation as
$$y'' + n^2 y = \frac{W}{2EI}(x^2 - \ell x) \quad \text{where} \quad \frac{P}{EI} = n^2,$$
C.F. is $A \cos nx + B \sin nx$.

For P.I. try $y = \alpha x^2 + \beta x + \gamma$ and substitution gives
$$2\alpha + n^2(\alpha x^2 + \beta x + \gamma) \equiv \frac{Wn^2}{2P}(x^2 - \ell x).$$

This results in $\alpha = \frac{W}{2P}$, $\beta = -\frac{W\ell}{2P}$, and $\gamma = -\frac{W}{n^2 P}$

∴ G.S. is $y = A \cos nx + B \sin nx + \frac{W}{2P}x^2 - \frac{W\ell}{2P}x - \frac{W}{n^2 P}$.

$y = 0$ at $x = 0$ gives $A = \frac{W}{n^2 P}$,

$y = 0$ at $x = \ell$ gives $B = \frac{W}{n^2 P}\left(\frac{1 - \cos n\ell}{\sin n\ell}\right) = \frac{W}{n^2 P} \tan \frac{n\ell}{2}$,

∴ $y = \frac{W}{n^2 P}(\cos nx + \tan \frac{n\ell}{2} \sin nx) + \frac{W}{2P}(x^2 - \ell x - \frac{2}{n^2})$.

When $x = \frac{\ell}{2}$, manipulation yields
$$y = \frac{W}{n^2 P}\left(\sec \frac{n\ell}{2} - 1 - \frac{n^2 \ell^2}{8}\right).$$

DIFFERENTIAL EQUATIONS WITH CONSTANT COEFFICIENTS

– SOLUTION BY D-OPERATOR METHODS

A PROGRAMMED TEXT

A. C. Bajpai
I. M. Calus

INSTRUCTIONS

This programme constitutes a self-instructional course on the solution of differential equations with constant coefficients by D-operator methods.

The programme is divided up into a number of FRAMES which are to be worked *in the order given*. You will be required to participate in many of these frames and in such cases the answers are provided in ANSWER FRAMES, designated by the letter A following the frame number. Steps in the working are given where this is considered helpful. The answer frame is separated from the main frame by a line of asterisks: ******************. Keep the answers covered until you have written your own response. If your answer is wrong, go back and try to see why. Do not proceed to the next frame until you have corrected any mistakes in your attempt and are satisfied that you understand the contents up to this point.

Before reading this programme it is necessary that you are familiar with the following

Prerequisites

Binomial Theorem.

Complex numbers — algebraic manipulation, equating real and imaginary parts, exponential form.

The contents of Programme 2 as far as FRAME 28. It would be helpful to be familiar with the remainder of it, but this is not absolutely necessary.

CONTENTS

Instructions

FRAMES

1 - 2	Introduction
3	Definition
4 - 7	Fundamental Laws
8	Inverse Operator
9 - 13	Operation of $f(D)$ on certain functions
14 - 25	Finding the Particular Integral
26 - 27	General Solutions
28 - 33	Particular Integral - Cases of Failure
34	Summary
35	Miscellaneous examples
36	Answers to miscellaneous examples

SOLUTION BY D-OPERATOR METHODS

FRAME 1

Introduction

The general solution of the differential equation (d.e.)
$$ay'' + by' + cy = Q \qquad (1.1)$$
where a, b and c are constants, and Q is a function of x only, takes the form
$$y = C.F. + P.I.$$

You will remember that in a second order d.e. the Complementary Function (C.F.) contains two arbitrary constants and the Particular Integral (P.I.) contains no arbitrary constant.

(For reference, see FRAME 15 of the programme "Second Order Differential Equations with Constant Coefficients — Solution by Trial Methods".)

FRAME 2

We will assume that you are familiar with the determination of the C.F. In this programme we will develop D-operator techniques for obtaining the P.I. for various forms of Q*. The algebra involved in trial methods of solution becomes at times rather tedious. Standard rules of procedure based on the principles of differentiation can be used to simplify the working.

* i.e. functions of x on the R.H.S. of the d.e. with constant coefficients.

FRAME 3

Definition

The symbol D is used to denote differentiation with respect to the independent variable e.g. when $y = f(x)$, Dy represents $\frac{dy}{dx}$ or $D \equiv \frac{d}{dx}$. D is said to 'operate' on y giving $\frac{dy}{dx}$.

Thus $D(Dy)$ means $\frac{d}{dx}\left(\frac{dy}{dx}\right)$ i.e. $\frac{d^2y}{dx^2}$. This is written as D^2y.

Now express $\frac{d^3y}{dx^3}$ and $\frac{d^ny}{dx^n}$ using the D-operator.

****

3A

$$\frac{d^3y}{dx^3} = \frac{d}{dx}\left(\frac{d^2y}{dx^2}\right) = D(D^2y) = D^3y$$

$$\frac{d^ny}{dx^n} = D^ny$$

FRAME 4

Fundamental Laws

We shall now obtain the fundamental laws concerning the D-operator.

Let u and v be functions of x.

Then $\frac{d}{dx}(u + v) = \frac{d}{dx}u + \frac{d}{dx}v$.

Replacing the $\frac{d}{dx}$ by D, this becomes

$$D(u + v) = Du + Dv \qquad \text{DISTRIBUTIVE LAW}$$

This can be extended to $D(au + bv) = aDu + bDv$ where a and b are <u>constants</u>.

FRAME 5

You know that

$$\frac{d^p}{dx^p}\left(\frac{d^q}{dx^q}y\right) = \frac{d^{p+q}}{dx^{p+q}}y$$

and

$$\frac{d^q}{dx^q}\left(\frac{d^p}{dx^p}y\right) = \frac{d^{q+p}}{dx^{q+p}}y \quad \text{where p and q are positive integers}$$

giving $\quad D^p(D^qy) = D^q(D^py) \qquad$ COMMUTATIVE LAW

and $\quad D^p(D^qy) = D^{p+q}y \qquad$ INDEX LAW

SOLUTION BY D-OPERATOR METHODS

FRAME 6

We can write $\dfrac{dy}{dx} + ay$ as $Dy + ay$ i.e. $(D + a)y$ and treat $(D + a)$ as an operator.

Write the following in a similar form, i.e. $f(D)y$:

(i) $\quad \dfrac{d^2y}{dx^2} + 9y$

(ii) $\quad \dfrac{d^2y}{dx^2} + 2\dfrac{dy}{dx} - 3y$

(iii) $\quad \dfrac{d^4y}{dx^4} - k^4y$

****

6A

(i) $\quad (D^2 + 9)y$

(ii) $\quad (D^2 + 2D - 3)y$

(iii) $\quad (D^4 - k^4)y$

FRAME 7

Extending the concept of $(D + a)$ as an operator, we can treat $(D + a)(D + b)$ also as an operator so that

$$(D + a)(D + b)y = (D + a)(Dy + by)$$
$$= D(Dy + by) + a(Dy + by)$$
$$= D^2y + (a + b)Dy + aby$$

which is the same as $\{D^2 + (a + b)D + ab\}y$, and

$(D + b)(D + a)y$ will give the same result.

You will note that

$$D^2 + (a + b)D + ab \equiv (D + a)(D + b) \equiv (D + b)(D + a).$$

DIFFERENTIAL EQUATIONS WITH CONSTANT COEFFICIENTS

FRAME 7 continued

From this we conclude that if f(D), where f(D) is a polynomial in D, operates on y, then

    (i)   f(D) can be factorised

    (ii)   the order of the factors is immaterial.

Now go back to the answers in 6A and express them in factor form.

****

7A

(i)   *(D + 3i)(D − 3i)y*   or   *(D − 3i)(D + 3i)y*

(ii)   *(D + 3)(D − 1)y*   or   *(D − 1)(D + 3)y*

(iii)   *One of the possible forms is* *(D − ki)(D + ki)(D − k)(D + k)y*

FRAME 8

Inverse Operator

The inverse operator D^{-1} must be such that

$$D^{-1}(Dy) = y$$

i.e. $D(D^{-1}y) = y$

or $\frac{d}{dx}(D^{-1}y) = y$

which, on integration, gives $D^{-1}y = \int y\,dx.$

You will note that the inverse operator is, in fact, integration. The arbitrary constant need not be included when finding the P.I.

Find   (i) $(D^{-1} + 1 + \frac{3D}{2})x^2$

             (ii) $\frac{1}{D^2}e^{3x}$

****

8A

(i) $\frac{x^3}{3} + x^2 + 3x$

(ii) $D^{-2}e^{3x} = D^{-1}(D^{-1}e^{3x}) = D^{-1}\frac{1}{3}e^{3x}$

                                                 $= \frac{1}{9}e^{3x}$

SOLUTION BY D-OPERATOR METHODS

FRAME 9

Operation of f(D) on certain functions

Let us now look at the effect of the D-operator on

(a) $e^{\alpha x}$

(b) $\sin \alpha x$ or $\cos \alpha x$

(c) $e^{\alpha x} V$ where V is a function of x.

(α is a constant in all cases.)

Consider (a). The rule of differentiation gives

$$De^{\alpha x} = \alpha e^{\alpha x}$$

$$(D + a)e^{\alpha x} = De^{\alpha x} + ae^{\alpha x}$$

$$= (\alpha + a)e^{\alpha x}$$

Write down the answers to the following:

$D^2 e^{\alpha x}$, $\quad D^3 e^{\alpha x}$, $\quad D^n e^{\alpha x}$, $\quad (D^2 + a^2)e^{\alpha x}$, $\quad (D^2 + 2D - 5)e^{\alpha x}$

****

9A

$\alpha^2 e^{\alpha x}$, $\quad \alpha^3 e^{\alpha x}$, $\quad \alpha^n e^{\alpha x}$, $\quad (\alpha^2 + a^2)e^{\alpha x}$, $\quad (\alpha^2 + 2\alpha - 5)e^{\alpha x}$

FRAME 10

From the answers in 9A you will notice that with the function $e^{\alpha x}$, D is replaced by α.

This gives us the standard result: $\quad f(D)e^{\alpha x} = f(\alpha)e^{\alpha x} \quad$ (10.1)
where f(D) is a polynomial in D.

Using this result, we get, for example, $(D + 1)^2 e^{-3x} = (-3 + 1)^2 e^{-3x}$

$$= 4e^{-3x}.$$

Now obtain the following:

$$(2D^3 - 3D^2 + 1)e^{-x}, \quad (D - 2)(D + 1)e^{4x}, \quad (D^2 + 8)e^{2ix}$$

****

10A

$$-4e^{-x}, \qquad 2.5e^{4x} = 10e^{4x}, \qquad 4e^{2ix}$$

FRAME 11

Proceeding now to (b) i.e. $\sin \alpha x$ or $\cos \alpha x$, the rule of differentiation gives

$$D \sin \alpha x = \alpha \cos \alpha x \quad \text{and} \quad D^2 \sin \alpha x = (-\alpha^2) \sin \alpha x$$

also, $\quad D \cos \alpha x = -\alpha \sin \alpha x \quad \text{and} \quad D^2 \cos \alpha x = (-\alpha^2) \cos \alpha x.$

You will note that we get a multiple of the original function only after differentiating <u>twice</u> (in general, an <u>even</u> number of times).

This property gives us the standard results:

$$\left. \begin{array}{l} f(D^2) \sin \alpha x = f(-\alpha^2) \sin \alpha x \\ f(D^2) \cos \alpha x = f(-\alpha^2) \cos \alpha x \end{array} \right\} \qquad (11.1)$$

As an illustration, $(D^4 + D^2 - 7) \sin \alpha x = (\alpha^4 - \alpha^2 - 7) \sin \alpha x.$
(D^2 is replaced by $-\alpha^2$ and hence D^4 is replaced by α^4.)
Odd powers of D will be dealt with in a later frame.

Obtain the following:

$$(D^4 - D^2 + 3) \cos \alpha x, \quad (D^2 + a^2)^2 \sin bx, \quad (D^4 - k^4) \cos ax$$

****

11A

$$(\alpha^4 + \alpha^2 + 3) \cos \alpha x, \quad (a^2 - b^2)^2 \sin bx, \quad (a^4 - k^4) \cos ax$$

FRAME 12

For further practice obtain:

(i) $\quad (D^2 - 4)(2e^{-x} + \sin 3x)$

(ii) $\quad (D^2 - 5D + 6) \cosh 2x \qquad \{\text{HINT: } \cosh \alpha x = \tfrac{1}{2}(e^{\alpha x} + e^{-\alpha x})\}$

(iii) $\quad (D^2 - 2D + 1) \sinh \dfrac{x}{3}$

****

SOLUTION BY D-OPERATOR METHODS

12A

(i) $2\{(-1)^2 - 4\}e^{-x} + (-3^2 - 4)\sin 3x = -6e^{-x} - 13\sin 3x$

(ii) $\tfrac{1}{2}\{(4 - 10 + 6)e^{2x} + (4 + 10 + 6)e^{-2x}\} = 10e^{-2x}$

Alternatively,
$$(D - 2)(D - 3)\{\tfrac{1}{2}(e^{2x} + e^{-2x})\} = 10e^{-2x}$$

(iii) $\tfrac{2}{9}(e^{x/3} - 4e^{-x/3})$ NOTE: $f(D)$ factorises.

FRAME 13

Let us now consider the operation of $f(D)$ on (c) of FRAME 9, i.e. $e^{\alpha x}V$, where V is a function of x.

$$\begin{aligned}
D(e^{\alpha x}V) &= \alpha e^{\alpha x}V + e^{\alpha x}DV \quad \text{by Product Rule} \\
&= e^{\alpha x}(D + \alpha)V
\end{aligned}$$

$$\begin{aligned}
D^2(e^{\alpha x}V) &= D\{e^{\alpha x}(D + \alpha)V\} \\
&= D(e^{\alpha x}U) \quad \text{where } U = (D + \alpha)V \\
&= e^{\alpha x}(D + \alpha)U \quad \text{by previous result} \\
&= e^{\alpha x}(D + \alpha)^2 V
\end{aligned}$$

Repetition of the above will give the general result:

$$D^m(e^{\alpha x}V) = e^{\alpha x}(D + \alpha)^m V \quad \text{where } m \text{ is a positive integer.}$$

This leads to the standard result:

$$f(D)(e^{\alpha x}V) = e^{\alpha x}f(D + \alpha)V \qquad (13.1)$$

i.e. every D in the polynomial $f(D)$ is replaced by $(D + \alpha)$ to give $f(D + \alpha)$. This is often called the "Shift Theorem".

As an example,
$$\begin{aligned}
(3D^2 + 5D - 7)e^{2x}x^2 &= e^{2x}\{3(D + 2)^2 + 5(D + 2) - 7\}x^2 \\
&= e^{2x}(3D^2 + 17D + 15)x^2 \\
&= e^{2x}(6 + 34x + 15x^2).
\end{aligned}$$

Now try the following:

(i) $(2D^2 - D + 3)e^{-x}\sin x$

(ii) $(D^2 - 2D - 3)xe^{3x}$ HINT: $f(D)$ factorises.

****

(i) $e^{-x}(4 \sin x - 5 \cos x)$

(ii) $4e^{3x}$

___

FRAME 14

Finding the Particular Integral

The d.e. (1.1) can be written in the form $(aD^2 + bD + c)y = Q$. The general d.e. of order n with constant coefficients is

$$(a_n D^n + a_{n-1} D^{n-1} + \ldots + a_1 D + a_0)y = Q.$$

This can be written as

$$f(D)y = Q, \qquad (14.1)$$

where $f(D)$ is a polynomial of degree n.

The solution of $f(D)y = 0$ gives the C.F. with n arbitrary constants.

To obtain the P.I.:

Let $\frac{1}{f(D)} Q$ be a function of x such that

$$f(D) \left[\frac{1}{f(D)} Q \right] = Q. \qquad (14.2)$$

Then $y = \frac{1}{f(D)} Q$ is obviously a particular solution of (14.1).

You will note that $\frac{1}{f(D)}$ here is an operator, operating on Q, yielding the P.I.

___

FRAME 15

We shall now consider how to obtain P.I.'s for various forms of Q commonly met with in problems from science and engineering.

These forms of Q are:

(i) polynomial expressions e.g. $\alpha x^2 + \beta x + \gamma$

(ii) $k e^{\alpha x}$

(iii) $k \sin \alpha x$ or $k \cos \alpha x$

(iv) $k e^{\alpha x} V$ where V is a function of x

k, α, β and γ are constants.

SOLUTION BY D-OPERATOR METHODS

FRAME 16

Case (i): **Polynomial expressions** e.g. $\alpha x^2 + \beta x + \gamma$

To operate with $\dfrac{1}{f(D)}$ on powers of x, we expand the operator $\dfrac{1}{f(D)}$ in ascending powers of D by using the Binomial Theorem.

The following example illustrates the procedure:

Consider the d.e.
$$(D^2 - 3D + 2)y = x^2 + 3x - 4.$$

The P.I. is given by
$$y = \frac{1}{D^2 - 3D + 2}(x^2 + 3x - 4)$$

Considering the operator only, this can be written as

$$\frac{1}{2\left(1 - \frac{3D}{2} + \frac{D^2}{2}\right)} = \frac{1}{2}\left\{1 + \left(\frac{-3D}{2} + \frac{D^2}{2}\right)\right\}^{-1}$$

$$= \frac{1}{2}\left\{1 - \left(\frac{-3D}{2} + \frac{D^2}{2}\right) + \left(\frac{-3D}{2} + \frac{D^2}{2}\right)^2 \ldots\right\}$$

using the Binomial Theorem.

Note that the expansion is only required as far as D^2 in this example, as differentiating more than twice gives zero.

Hence $y = \dfrac{1}{2}\left(1 + \dfrac{3D}{2} - \dfrac{D^2}{2} + \dfrac{9D^2}{4} \ldots\right)(x^2 + 3x - 4)$

$= \dfrac{1}{2}\left(1 + \dfrac{3D}{2} + \dfrac{7D^2}{4}\right)(x^2 + 3x - 4)$ (16.1)

$= \dfrac{1}{2}\{x^2 + 3x - 4 + \dfrac{3}{2}(2x + 3) + \dfrac{7}{4} \cdot 2\}$

$= \dfrac{1}{2}(x^2 + 6x + 4),$ which is the required P.I.

FRAME 17

Alternatively, the method of partial fractions could be used to obtain the expansion in cases where f(D) has factors.

Express $\dfrac{1}{D^2 - 3D + 2}$ in partial fractions
and hence expand as far as D^2 to obtain it in the form

$$\frac{1}{2}\left(1 + \frac{3D}{2} + \frac{7D^2}{4}\right), \quad \text{as in (16.1)}.$$

***

17A

$$\begin{aligned}
\frac{1}{(D-1)(D-2)} &= -\frac{1}{D-1} + \frac{1}{D-2} \\
&= (1-D)^{-1} - \frac{1}{2}\left(1 - \frac{D}{2}\right)^{-1} \\
&= 1 + D + D^2 - \frac{1}{2}\left(1 + \frac{D}{2} + \frac{D^2}{4}\right) \\
&= \frac{1}{2}\left(1 + \frac{3D}{2} + \frac{7D^2}{4}\right)
\end{aligned}$$

FRAME 18

Now find the P.I. for each of the following d.e.'s:

(i) $(D^2 + D + 1)y = x^3 - 2$

(ii) $(D^2 - 3D)y = 2x + 3$

***

18A

(i) $x^3 - 3x^2 + 4$

(ii) $-\dfrac{1}{27}(9x^2 + 33x + 11)$

SOLUTION BY D-OPERATOR METHODS 3:11

FRAME 19

Case (ii): $ke^{\alpha x}$

You will recall that in FRAME 10 we obtained the standard result (10.1)
$$f(D)e^{\alpha x} = f(\alpha)e^{\alpha x}.$$
This suggests that $\dfrac{1}{f(D)}e^{\alpha x} = \dfrac{1}{f(\alpha)}e^{\alpha x}$ \qquad (19.1)
provided that $f(\alpha) \neq 0$.

The justification for this lies in the fact that operating with $f(D)$ on both sides of (19.1) yields $e^{\alpha x}$:

$$f(D)\left\{\frac{1}{f(D)}e^{\alpha x}\right\} = e^{\alpha x} \qquad \text{from (14.2)}$$

$$\text{and} \quad f(D)\left\{\frac{1}{f(\alpha)}e^{\alpha x}\right\} = \frac{1}{f(\alpha)}f(D)e^{\alpha x} \qquad \text{as } f(\alpha) \text{ is a constant}$$

$$= \frac{1}{f(\alpha)}f(\alpha)e^{\alpha x} \qquad \text{from (10.1)}$$

$$= e^{\alpha x}$$

The case when $f(\alpha) = 0$ will be discussed later.

We now use (19.1) to find the P.I. for the d.e.
$$(3D^2 - 5D - 4)y = 5e^{2x}.$$
The P.I. is given by $y = \dfrac{1}{3D^2 - 5D - 4} 5e^{2x}$

$$= 5 \cdot \frac{1}{3(2)^2 - 5(2) - 4} e^{2x} \qquad \text{NOTE: } f(2) \neq 0$$

$$= -\frac{5}{2} e^{2x}.$$

Now try the following examples:

(i) Find $\dfrac{1}{D^2 - 2D + 7} 2e^{-x}$

(ii) Find the P.I. for the d.e.
$(2D^2 - 3D + 2)y = 2 \cosh \dfrac{x}{2}$

****

19A

(i) $\dfrac{1}{5} e^{-x}$

(ii) $e^{x/2} + \dfrac{1}{4} e^{-x/2}$

DIFFERENTIAL EQUATIONS WITH CONSTANT COEFFICIENTS

FRAME 20

Case (iii): $k \sin \alpha x$ or $k \cos \alpha x$

In FRAME 11 we obtained the standard results (11.1)

$$f(D^2) \begin{array}{c} \sin \alpha x \\ \cos \alpha x \end{array} = f(-\alpha^2) \begin{array}{c} \sin \alpha x \\ \cos \alpha x \end{array}.$$

The top result suggests that $\dfrac{1}{f(D^2)} \sin \alpha x = \dfrac{1}{f(-\alpha^2)} \sin \alpha x$ \quad (20.1)

provided that $f(-\alpha^2) \neq 0$. The justification for this lies in the fact that operating with $f(D^2)$ on both sides of (20.1) yields $\sin \alpha x$.

It follows from (14.2) that

$$f(D^2)\left\{\frac{1}{f(D^2)} \sin \alpha x\right\} = \sin \alpha x,$$

also $\quad f(D^2)\left\{\dfrac{1}{f(-\alpha^2)} \sin \alpha x\right\} = \dfrac{1}{f(-\alpha^2)} f(D^2) \sin \alpha x$ \quad as $f(-\alpha^2)$ is constant

$$= \sin \alpha x \quad \text{from (11.1).}$$

Obviously a similar result is

$$\frac{1}{f(D^2)} \cos \alpha x = \frac{1}{f(-\alpha^2)} \cos \alpha x.$$

The case when $f(-\alpha^2) = 0$ will be discussed later.

We now use (20.1) to find the P.I. for the d.e.

$$(D^2 - 9)y = 5 \sin 4x.$$

The P.I. is given by

$$y = \frac{1}{D^2 - 9} 5 \sin 4x$$

$$= 5 \cdot \frac{1}{-(4)^2 - 9} \sin 4x \qquad \text{NOTE: } f(-4^2) \neq 0$$

$$= -\frac{1}{5} \sin 4x. \qquad (20.2)$$

Now find $\dfrac{1}{2D^2 + 1} 7 \cos 5x$.

****

20A

$$-\frac{1}{7} \cos 5x$$

SOLUTION BY D-OPERATOR METHODS

FRAME 21

You will recall that in FRAME 11 we did not consider polynomials in D including <u>odd</u> powers of D. We shall illustrate the technique for dealing with such cases by examples.

Consider
$$\frac{1}{D-3} 5 \sin 4x. \qquad (21.1)$$

Since the standard rule of procedure involves $\frac{1}{f(D^2)}$ we shall first express the operator in (21.1) in this form by multiplying both numerator and denominator by $(D + 3)$.

This gives

$$\frac{D+3}{D^2-9} 5 \sin 4x = (D+3)\frac{1}{D^2-9} 5 \sin 4x$$

$$= (D+3)(-\frac{1}{5} \sin 4x) \qquad \text{from (20.2)}$$

$$= -\frac{1}{5}(4 \cos 4x + 3 \sin 4x).$$

Now you should try $\frac{1}{D+2} \cos 3x$.

****

21A

$\frac{1}{13}(3 \sin 3x + 2 \cos 3x)$

FRAME 22

Let us now consider in the next example the case of $f(D)$ containing both odd and even powers of D.

$$\frac{1}{D^2 - 2D + 19} \sin 4x = \frac{1}{-(4)^2 - 2D + 19} \sin 4x$$

$$= \frac{1}{-2D + 3} \sin 4x$$

$$= \frac{-2D - 3}{4D^2 - 9} \sin 4x$$

$$= \frac{-(2D + 3)}{-64 - 9} \sin 4x$$

$$= \frac{1}{73}(8 \cos 4x + 3 \sin 4x)$$

Now obtain $\frac{1}{D^2 - 3D + 2} 13 \cos 3x$.

****

22A

$$-\frac{1}{10}(9 \sin 3x + 7 \cos 3x)$$

FRAME 23

You will recall that we obtained standard results (11.1) for $\sin \alpha x$ and $\cos \alpha x$. You will readily see that a similar result can be obtained for $\sinh \alpha x$ and $\cosh \alpha x$,

i.e. $\qquad f(D^2) \begin{array}{l} \sinh \alpha x \\ \cosh \alpha x \end{array} = f(\alpha^2) \begin{array}{l} \sinh \alpha x \\ \cosh \alpha x \end{array}$ \qquad (23.1)

Therefore, if Q were given as $k \sinh \alpha x$ or $k \cosh \alpha x$ it could be treated in a similar way to $k \sin \alpha x$ or $k \cos \alpha x$.

Perhaps you would like to use this method to find the P.I. for the d.e. in FRAME 19, example (ii), i.e. $(2D^2 - 3D + 2)y = 2 \cosh \frac{x}{2}$.

****

23A

$$y = \frac{1}{2D^2 - 3D + 2} \, 2 \cosh \frac{x}{2}$$

$$= 2 \cdot \frac{1}{\frac{5}{2} - 3D} \cosh \frac{x}{2} \quad \text{on replacing } D^2 \text{ by } (\tfrac{1}{2})^2$$

$$= 2 \cdot \frac{\frac{5}{2} + 3D}{\frac{25}{4} - 9D^2} \cosh \frac{x}{2}$$

$$= \frac{1}{4}(5 \cosh \frac{x}{2} + 3 \sinh \frac{x}{2})$$

This could be expressed in the form $e^{x/2} + \frac{1}{4} e^{-x/2}$ as in 19A (ii).

SOLUTION BY D-OPERATOR METHODS 3:15

FRAME 24

Case (iv): $ke^{\alpha x}V$, where V is a function of x

In FRAME 13 we obtained the standard result (13.1)

$$f(D)\{e^{\alpha x}V\} = e^{\alpha x}f(D + \alpha)V.$$

This suggests that $\dfrac{1}{f(D)}\{e^{\alpha x}V\} = e^{\alpha x}\dfrac{1}{f(D + \alpha)}V.$ (24.1)

The justification for this lies in the fact that operating with $f(D)$ on both sides of (24.1) yields $e^{\alpha x}V$,

for $\quad f(D)\left[\dfrac{1}{f(D)}\{e^{\alpha x}V\}\right] = e^{\alpha x}V \qquad$ from (14.2)

and $f(D)\left[e^{\alpha x}\{\dfrac{1}{f(D + \alpha)}V\}\right] = e^{\alpha x}f(D + \alpha)\{\dfrac{1}{f(D + \alpha)}V\} \qquad$ from (13.1)

$\qquad\qquad\qquad\qquad\qquad = e^{\alpha x}V \quad$ since $f(D + \alpha)$ and $\{f(D + \alpha)\}^{-1}$ are inverse operators.

We now use (24.1) to find the P.I. for the d.e.

$$(D^2 - 2D + 5)y = 10e^{2x}\sin x.$$

The P.I. is given by $y = \dfrac{1}{D^2 - 2D + 5} 10e^{2x}\sin x$

$\qquad\qquad\qquad = 10e^{2x}\dfrac{1}{(D + 2)^2 - 2(D + 2) + 5}\sin x$

$\qquad\qquad\qquad = 10e^{2x}\dfrac{1}{D^2 + 2D + 5}\sin x.$

Now complete the solution.

****

24A

$$y = 10e^{2x}\dfrac{1}{2D + 4}\sin x$$

$$= 5e^{2x}\dfrac{D - 2}{D^2 - 4}\sin x$$

$$= e^{2x}(2\sin x - \cos x)$$

FRAME 25

Now try the following examples:

(i) Find the P.I. for the d.e. $(3D^2 + 4D + 5)y = 20e^{-x}\cos 2x$

(ii) Obtain $\dfrac{1}{(D-1)^2} x^2 e^{3x}$

****

25A

(i) $-e^{-x}(\sin 2x + 2\cos 2x)$

(ii) $\dfrac{e^{3x}}{8}(2x^2 - 4x + 3)$

FRAME 26

In the previous frames you have seen how to obtain P.I.'s for the d.e.'s with various forms of Q, using D-operator techniques.

We shall now find the general solutions of two such d.e.'s, and in one instance, also the particular solution given by certain boundary conditions.

Example (i)

Solve the d.e. $y'' - 3y' + 2y = 13\cos 3x$ given that $y = y' = 0$ when $x = 0$. First find the C.F. Then write down the general solution - you have already obtained the P.I. in FRAME 22 - and apply the given boundary conditions to obtain the particular solution required.

****

26A

The A.E. is $m^2 - 3m + 2 = 0$ *giving the C.F. as* $Ae^{2x} + Be^{x}$.

The P.I. is $\dfrac{-1}{10}(9\sin 3x + 7\cos 3x)$ *from 22A.*

$\therefore$ *G.S. is* $y = Ae^{2x} + Be^{x} - \dfrac{1}{10}(9\sin 3x + 7\cos 3x)$

$x = 0, y = 0$ *gives* $A + B = \dfrac{7}{10}$

$x = 0, y' = 0$ *gives* $2A + B = \dfrac{27}{10}$

SOLUTION BY D-OPERATOR METHODS   3:17

                                                        26A continued

Solving for A and B, we get $A = 2$, $B = -\frac{13}{10}$.

The required particular solution is

$$y = 2e^{2x} - \frac{13}{10} e^{x} - \frac{1}{10}(9 \sin 3x + 7 \cos 3x).$$

                                                        FRAME 27

Example (ii)

Solve the d.e. $3 \frac{d^2 i}{dt^2} + 4 \frac{di}{dt} + 5i = 20 e^{-t} \cos 2t.$

(NOTE: You have already found the P.I. in FRAME 25, Example (i) with variables y and x. In the above d.e. $D \equiv \frac{d}{dt}$ as the independent variable is t, instead of x.)

****

                                                        27A

The d.e. can be written $(3D^2 + 4D + 5)i = 20 e^{-t} \cos 2t.$

The C.F. is $e^{-2t/3}(A \cos \frac{\sqrt{11}}{3} t + B \sin \frac{\sqrt{11}}{3} t).$

The P.I. is $-e^{-t}(\sin 2t + 2 \cos 2t).$

The G.S. is $i = e^{-2t/3}(A \cos \frac{\sqrt{11}}{3} t + B \sin \frac{\sqrt{11}}{3} t) - e^{-t}(\sin 2t + 2 \cos 2t).$

                                                        FRAME 28

Particular Integral - Cases of Failure

In certain d.e.'s, some of the terms in the C.F. appear in Q, the function on the R.H.S. Such cases are referred to as CASES OF FAILURE and special techniques for the P.I. are required.

(Cases of failure were discussed in FRAME 49 of the earlier programme "Second Order D.E.'s with Constant Coefficients - Solution by Trial Methods".)

FRAME 28 continued

The three possible forms of the C.F. are:

(i) $Ae^{m_1 x} + Be^{m_2 x}$
(ii) $(A + Bx)e^{nx}$
(iii) $e^{px}(A \cos qx + B \sin qx)$

Cases of failure will occur if Q is of the form:

In (i), either $ke^{m_1 x}$ or $ke^{m_2 x}$
In (ii), ke^{nx} (kxe^{nx} will not be a case of failure since multiplication by x was used to modify the solution. See FRAME 22 of the earlier programme mentioned above.)
In (iii), $k \sin qx$ or $k \cos qx$, when $p = 0$, and
$ke^{px} \sin qx$ or $ke^{px} \cos qx$, when $p \neq 0$.

FRAME 29

Let us first practise recognising cases of failure.

Consider the d.e. $y'' - y' - 2y = 5e^{2x}$ (29.1)

The A.E. is $m^2 - m - 2 = 0$

giving the C.F. $Ae^{2x} + Be^{-x}$.

We see that e^{2x} occurs both in the C.F. and in Q. Hence there is a case of failure.

Now recognise which of the following are cases of failure:

(i) $y'' + 4y = 3 \sin 2x$
(ii) $y'' + 4y = e^x \cos 2x$
(iii) $y'' - 4y' + 5y = 2e^{2x} \cos x$

****

29A

(i)   *Yes. C.F. is* $A \cos 2x + B \sin 2x$ *and* $\sin 2x$ *also occurs in Q.*
(ii)   *No. C.F. is* $A \cos 2x + B \sin 2x$ *but the function in Q is* $e^x \cos 2x$.
(iii)   *Yes. C.F. is* $e^{2x}(A \cos x + B \sin x)$ *and* $e^{2x} \cos x$ *also occurs in Q.*

SOLUTION BY D-OPERATOR METHODS

FRAME 30

The P.I. for the d.e. (29.1) is

$$\frac{1}{D^2 - D - 2} 5e^{2x} = 5 \frac{1}{D^2 - D - 2} e^{2x} \qquad (30.1)$$

If you apply the standard rule of procedure (19.1), you will get a zero denominator, i.e. $f(2) = 0$. The exception to the rule, $f(\alpha) = 0$, is, of course, the case of failure.

Now verify that the exception mentioned in (20.1), i.e. $f(-\alpha^2) = 0$, occurs in examples (i) and (iii) in FRAME 29.

****

30A

(i) $\quad \dfrac{1}{D^2 + 4} 3 \sin 2x = 3 \dfrac{1}{-4 + 4} \sin 2x$

(iii) $\quad \dfrac{1}{D^2 - 4D + 5} 2e^{2x} \cos x = 2e^{2x} \dfrac{1}{D^2 + 1} \cos x$

$\qquad\qquad\qquad\qquad\qquad\qquad = 2e^{2x} \dfrac{1}{-1 + 1} \cos x$

FRAME 31

The presence of a zero denominator causes a breakdown of the methods discussed before. We will now develop a special technique for dealing with this situation.

The forms of Q giving rise to cases of failure as described in FRAME 28 are essentially **exponential**. (Remember that $\sin qx$ and $\cos qx$ are the imaginary and real parts, respectively, of e^{iqx}.) Going back to (30.1):

The P.I. $5 \dfrac{1}{D^2 - D - 2} e^{2x}$ could be written as $5 \dfrac{1}{(D - 2)(D + 1)} e^{2x}$.

Note that the factor $(D - 2)$ makes the denominator vanish.

We can, however, operate with the $\dfrac{1}{D + 1}$ giving

$$5 \frac{1}{D - 2} \frac{1}{3} e^{2x} = \frac{5}{3} \frac{1}{D - 2} e^{2x}.$$

FRAME 31 continued

Let us now consider e^{2x} as $e^{2x} \cdot 1$ which is of the form $e^{\alpha x}V$, where $\alpha = 2$ and $V = 1$. We can then apply the standard result (24.1).

$$\frac{5}{3} \frac{1}{D-2} e^{2x} \cdot 1 = \frac{5}{3} e^{2x} \frac{1}{D+2-2} \cdot 1$$

$$= \frac{5}{3} e^{2x} \frac{1}{D} \cdot 1$$

$$= \frac{5}{3} e^{2x} x$$

$\therefore$ The P.I. is $\frac{5}{3} xe^{2x}$.

FRAME 32

We now consider answer (i) in 30A.

The P.I. $\frac{1}{D^2+4} 3 \sin 2x$ could be written as $\operatorname{Im}\left(\frac{1}{D^2+4} 3e^{2ix}\right)$

$$\frac{1}{D^2+4} 3e^{2ix} = 3 \frac{1}{(D-2i)(D+2i)} e^{2ix}$$

Now proceed as in the previous frame and take the imaginary part to obtain the P.I.

****

32A

$$\frac{3}{4i} \frac{1}{D-2i} e^{2ix} \cdot 1 = \frac{3}{4i} e^{2ix} \frac{1}{D} \cdot 1$$

$$= \frac{3x}{4i} e^{2ix}$$

$$= -\frac{3xi}{4}(\cos 2x + i \sin 2x)$$

Taking the imaginary part, the P.I. is $-\frac{3}{4} x \cos 2x$.

SOLUTION BY D-OPERATOR METHODS 3:21

FRAME 33

Let us now see how the same technique is applied to answer (iii) in 30A.

We already have the P.I. $\dfrac{1}{D^2 - 4D + 5} 2e^{2x} \cos x = 2e^{2x} \dfrac{1}{D^2 + 1} \cos x$.

Using a similar procedure to that in FRAME 32, write $\cos x$ as the real part of an exponential function to obtain the P.I.

****

33A

$$P.I. = 2e^{2x} \, Re\left(\dfrac{1}{D^2 + 1} e^{ix}\right)$$

$$\dfrac{1}{D^2 + 1} e^{ix} = \dfrac{1}{(D - i)(D + i)} e^{ix}$$

$$= \dfrac{1}{2i} \dfrac{1}{D - i} e^{ix} . 1$$

$$= \dfrac{1}{2i} e^{ix} \dfrac{1}{D} . 1$$

$$= -\tfrac{1}{2} i x e^{ix}$$

$$= -\tfrac{1}{2} i x (\cos x + i \sin x)$$

$$P.I. = x e^{2x} \sin x.$$

FRAME 34

Summary

Before trying some miscellaneous examples you may find it useful to refer to the following summary of the various methods of obtaining P.I.'s using the D-operator.

The P.I. is given by $\dfrac{1}{f(D)} Q$.

<u>Case (i)</u>   Q a polynomial expression

Expand $\dfrac{1}{f(D)}$ in ascending powers of D.

FRAME 34 continued

Case (ii) $Q = ke^{\alpha x}$

Replace D by α in $f(D)$, provided that $f(\alpha) \neq 0$.

Case (iii) $Q = k \sin \alpha x$ or $k \cos \alpha x$

For $\dfrac{1}{f(D^2)}$ (even powers of D only) replace D^2 by $-\alpha^2$ provided that $f(-\alpha^2) \neq 0$.

For $\dfrac{1}{f(D)}$ (including odd powers of D) replace D^2 by $-\alpha^2$ as a first step. Then convert any D's remaining in the denominator to D^2 by multiplying both numerator and denominator by the appropriate factor. Replace D^2 by $-\alpha^2$ again and operate with the numerator on Q.

Case (iv) $Q = ke^{\alpha x}V$

Take $ke^{\alpha x}$ out as a factor and replace D by $(D + \alpha)$ to get $ke^{\alpha x} \dfrac{1}{f(D + \alpha)} V$. Then apply the technique appropriate to the function V.

Cases of failure

Q must be in <u>exponential</u> form (write $\sin qx = \text{Im } e^{iqx}$ and $\cos qx = \text{Re } e^{iqx}$). The technique of Case (iv) is applied with $V = 1$.
For details, see FRAMES 31 - 33.

FRAME 35

Miscellaneous Examples

In this frame, a collection of miscellaneous examples is given for you to try. Answers are supplied in FRAME 36 and hints have been provided in some cases.

Solve the following d.e.'s, using D-operator methods, subject to the conditions given where appropriate:

1. $4y'' - 4y' + y = x^2 - 3x - 13$

2. $y'' + 4y' + 4y = 2e^x - 3 \cos x$

3. $y'' + 4y' + 5y = 3e^{-2x}$ given that $y = 4$ and $y' = -7$ when $x = 0$.

4. $y'' - 3y' + 2y = \tfrac{1}{2}e^x$ given that $y = 0$ and $y' = 0$ when $x = 0$.

SOLUTION BY D-OPERATOR METHODS   3:23

FRAME 35 continued

5. $y'' - y = x \sin x$

6. $\dfrac{d^4 y}{dx^4} - 16y = 4e^{-2x} \cos 2x$

7. $y'' - 2y' + 5y = e^{-x} \sin 2x + e^x \sin 2x$

8. $y'' + 9y = 6 \cos 3x$ subject to $y = \dfrac{\pi}{4}$ when $x = 0$ and $y = \dfrac{\pi}{3}$ when $x = \dfrac{\pi}{6}$.

9. In an oscillator circuit the d.e. for the charge q on the condenser at time t is $L\ddot{q} + \dfrac{q}{C} = E_o \cos pt$ where L, C, E_o and p are constants. Find q and the current $i (= \dot{q})$ at time t, if both q and i are initially zero, in the cases:

    (i) $p \neq \omega$

    (ii) $p = \omega$

where $\omega^2 = 1/LC$.

10. A body of unit mass moves along the x-axis under the action of an attractive force of magnitude $\omega^2 x$ directed towards the origin O. It is subject also to a frictional resistance of magnitude $2k\dot{x}$ and to a driving force $e^{-kt} \cos pt$. The equation of motion may, therefore, be written as

$$\ddot{x} + 2k\dot{x} + \omega^2 x = e^{-kt} \cos pt.$$

If the resistance is sufficiently small, so that $k < \omega$, we may write $\omega^2 - k^2 = n^2$, where n is real. Obtain the solution of the equation when

    (i) $p \neq n$

    (ii) $p = n$,

assuming in both cases that the body starts from rest at the origin.

FRAME 36

Answers to Miscellaneous Examples

1. $y = (A + Bx)e^{\frac{1}{2}x} + x^2 + 5x - 1$

2. $y = (A + Bx)e^{-2x} + \dfrac{2}{9}e^x - \dfrac{3}{25}(3 \cos x + 4 \sin x)$

FRAME 36 continued

3. $y = e^{-2x}(\cos x + \sin x) + 3e^{-2x}$

4. $y = \tfrac{1}{2}(e^{2x} - e^x - xe^x)$

5. $y = Ae^x + Be^{-x} - \tfrac{1}{2}(\cos x + x \sin x)$

   $\{$Take $Q = x \sin x = \text{Im}(xe^{ix})\}$

6. A.E. is $m^4 - 16 = 0$ giving $m = \pm 2, \pm 2i$

   C.F. is $Ae^{2x} + Be^{-2x} + C \cos 2x + E \sin 2x$

   P.I. is $\dfrac{1}{D^4 - 16} 4e^{-2x}\cos 2x = 4e^{-2x} \dfrac{1}{(D-2)^4 - 16} \cos 2x$

   $= 4e^{-2x} \dfrac{1}{(D^2 - 4D)(D^2 - 4D + 8)} \cos 2x$

   $= -\dfrac{1}{4} e^{-2x} \dfrac{1}{(1 - D)(1 + D)} \cos 2x$

   $= -\dfrac{1}{20} e^{-2x}\cos 2x$

   G.S. is $y = Ae^{2x} + Be^{-2x} + C \cos 2x + E \sin 2x - \dfrac{1}{20} e^{-x}\cos 2x$

7. $y = e^x(A \cos 2x + B \sin 2x) + \dfrac{1}{20} e^{-x}(\sin 2x + 2 \cos 2x) - \dfrac{1}{4} xe^x\cos 2x$

8. $y = \dfrac{\pi}{4} \cos 3x + \dfrac{\pi}{6} \sin 3x + x \sin 3x$

9. (i) G.S. is $q = A \cos \omega t + B \sin \omega t + \dfrac{E_o}{L(\omega^2 - p^2)} \cos pt$

   and $i = \dot{q} = -A\omega \sin \omega t + B\omega \cos \omega t - \dfrac{E_o p}{L(\omega^2 - p^2)} \sin pt$

   Applying $q = 0 = i$ at $t = 0$ gives

   $q = \dfrac{E_o}{L(\omega^2 - p^2)} (\cos pt - \cos \omega t)$

   $i = \dfrac{E_o}{L(\omega^2 - p^2)} (\omega \sin \omega t - p \sin pt)$

SOLUTION BY D-OPERATOR METHODS 3:25

FRAME 36 continued

9. (ii) G.S. is $q = A \cos \omega t + B \sin \omega t + \dfrac{E_o}{2\omega L} t \sin \omega t$

and $\therefore i = -A\omega \sin \omega t + B\omega \cos \omega t + \dfrac{E_o}{2\omega L}(\sin \omega t + \omega t \cos \omega t)$

Applying $q = 0 = i$ at $t = 0$ gives

$$q = \dfrac{E_o}{2\omega L} t \sin \omega t$$

$$i = \dfrac{E_o}{2\omega L}(\sin \omega t + \omega t \cos \omega t).$$

10. (i) G.S. is $x = e^{-kt}\left(A \cos nt + B \sin nt + \dfrac{\cos pt}{n^2 - p^2}\right)$

Applying $x = 0$, $\dot{x} = 0$ at $t = 0$ gives

$$x = \dfrac{e^{-kt}}{n^2 - p^2}(\cos pt - \cos nt).$$

(ii) G.S. is $x = e^{-kt}\left(A \cos nt + B \sin nt + \dfrac{t}{2n} \sin nt\right)$

Applying $x = 0$, $\dot{x} = 0$ at $t = 0$ gives

$$x = \dfrac{t}{2n} e^{-kt} \sin nt.$$

DIFFERENTIAL EQUATIONS WITH CONSTANT COEFFICIENTS

— SOLUTION BY LAPLACE TRANSFORM METHODS

A PROGRAMMED TEXT

A. C. Bajpai
I. M. Calus

INSTRUCTIONS

This programme constitutes a self-instructional course on the solution of linear differential equations with constant coefficients by Laplace Transform methods.

The programme is divided up into a number of FRAMES which are to be worked *in the order given*. You will be required to participate in many of these frames and in such cases the answers are provided in ANSWER FRAMES, designated by the letter A following the frame number. Steps in the working are given where this is considered helpful. The answer frame is separated from the main frame by a line of asterisks: ******************. Keep the answers covered until you have written your own response. If your answer is wrong, go back and try to see why. Do not proceed to the next frame until you have corrected any mistakes in your attempt and are satisfied that you understand the contents up to this point.

Before reading this programme it is necessary that you are familiar with the following

Prerequisites

Partial fractions.

Complex numbers — algebraic manipulation, equating real and imaginary parts, exponential form.

Integration, including integration by parts and the use of reduction formulae.

CONTENTS

Instructions

FRAMES

1 - 3	Introduction
4 - 6	Definition
7	Inverse Laplace Transform
8	Alternative notations
9 - 10	Transform of t^n
11,14,15,19	Some properties of the Laplace Transform
12 - 13	Transforms of further functions
16 - 18	Use of the table of transforms
20 - 23	Use of partial fractions in finding inverse transforms
24 - 25	Transforms of derivatives
26 - 34	Solution of differential equations
35 - 38	Applications
39	Miscellaneous examples
40	Answers to miscellaneous examples
APPENDIX	Table of Laplace Transform pairs.

SOLUTION BY LAPLACE TRANSFORM METHODS

FRAME 1

Introduction

The solution of linear differential equations with constant coefficients by trial and D-operator methods has been dealt with in previous programmes. In this programme we shall consider the solution of such differential equations, with given initial conditions, by Laplace Transform methods.

In the following frames we shall attempt to indicate certain advantages of Laplace Transform methods over the methods already discussed.

FRAME 2

By using the LAPLACE TRANSFORMATION, a d.e. such as

$$a \frac{d^2y}{dx^2} + b \frac{dy}{dx} + cy = f(x), \quad \text{where } a, b, c \text{ are constants,}$$

can be "transformed" to an algebraic equation incorporating the given initial conditions. This equation is obtained by using a standard table, and the problem then becomes one of simple algebraic manipulation. The answer is found by again using the table which gives the "inverse transform".

This is analogous to the use of logarithms for multiplication and division in arithmetic. By the use of log tables the problem is "transformed" to one of addition or subtraction, and the answer is found by referring to the antilog tables.

FRAME 3

Furthermore, you will recall that, in the trial and D-operator methods, a general solution of the d.e. was found before the initial conditions could be applied to obtain the arbitrary constants. This is sometimes a tedious process. Laplace Transform methods incorporate the initial conditions in the solution at the outset. This will be clearer to you later in the programme.

It is worth mentioning that in the special case of electrical engineering transient and steady state problems there are advantages in changing over from the time domain to the frequency domain.

Definition

In FRAME 2 we referred to a d.e in which x is the independent variable. In many engineering problems, functions of <u>time</u>, f(t), say, are involved so we shall use t as the independent variable, for convenience. The LAPLACE TRANSFORM of a function f(t) is defined as the integral

$$\int_0^\infty e^{-st} f(t) \, dt \qquad (4.1)$$

where s is a parameter (real or complex) such that the integral exists when $t \to \infty$. This imposes certain restrictions on the real part of s as you will see in the following answer frame. In this programme we shall confine ourselves to real values of s only. Now use the definition in (4.1) to obtain the Laplace Transform (L.T.) of $e^{\alpha t}$, where α is a constant.

***

<u>4A</u>

$$f(t) = e^{\alpha t}$$

$$\therefore \text{L.T. of } e^{\alpha t} = \int_0^\infty e^{-st} e^{\alpha t} \, dt$$

$$= \int_0^\infty e^{-(s-\alpha)t} \, dt$$

$$= -\frac{1}{s-\alpha} \left[e^{-(s-\alpha)t} \right]_0^\infty$$

$$= -\frac{1}{s-\alpha}(0 - 1)$$

$$= \frac{1}{s-\alpha} \qquad (4A.1)$$

Note that the integrand was written as $e^{-(s-\alpha)t}$. This is preferable as it is then clear that $e^{-(s-\alpha)t} \to 0$ as $t \to \infty$, because the index is negative, provided that $s > \alpha$.

– SOLUTION BY LAPLACE TRANSFORM METHODS

FRAME 5

When you found the L.T. of $e^{\alpha t}$ in the last example, you obtained $\frac{1}{s-\alpha}$ which is a function of s. Before drawing any general conclusions, find the L.T. for the case of $f(t) = 1$.

****

5A

$$f(t) = 1$$

$$\text{L.T. of } 1 = \int_0^\infty e^{-st} \cdot 1 \, dt$$

$$= -\frac{1}{s}\left[e^{-st}\right]_0^\infty$$

$$= -\frac{1}{s}(0 - 1) \quad \text{provided that } s > 0$$

$$= \frac{1}{s} \qquad\qquad (5A.1)$$

FRAME 6

These two examples should make it obvious to you that the integral in (4.1)

$$\int_0^\infty e^{-st} f(t) \, dt$$

will always be a function of s, as was the case in (4A.1) and (5A.1). We shall write

$$F(s) = \int_0^\infty e^{-st} f(t) \, dt \qquad (6.1)$$

and we say that F(s) is the Laplace Transform of f(t).

$\mathcal{L}$ is used to denote "the Laplace Transform of", so that

$$\mathcal{L}\{f(t)\} = F(s) \qquad (6.2)$$

f(t) and F(s) are "transform pairs", i.e. each f(t) has a corresponding F(s), using the small and capital form of the same letter to represent this relationship. The results already obtained could be tabulated as follows:

f(t)	F(s)
$e^{\alpha t}$	$\frac{1}{s-\alpha}$
1	$\frac{1}{s}$

FRAME 7

Inverse Laplace Transform

From the table in the previous frame we see that $e^{\alpha t}$ is the function whose transform is $\dfrac{1}{s - \alpha}$, i.e. $e^{\alpha t}$ is the inverse Laplace Transform of $\dfrac{1}{s - \alpha}$, usually written

$$\mathcal{L}^{-1}\left\{\frac{1}{s - \alpha}\right\} = e^{\alpha t}.$$

In general, $\mathcal{L}^{-1}\{F(s)\} = f(t)$.

You will note that the operation of the Laplace Transform on both sides yields

$$\mathcal{L}\left[\mathcal{L}^{-1}\{F(s)\}\right] = \mathcal{L}\{f(t)\}$$

i.e. $F(s) = \mathcal{L}\{f(t)\}$ as in (6.2).

You should appreciate that $\mathcal{L}\,\mathcal{L}^{-1} = 1$, as is the case in inverse operations.

FRAME 8

Alternative notations

We ought to mention, in passing, that sometimes other notations are used.

p is used in place of s

L is used in place of $\mathcal{L}$

$f(t) \supset F(s)$ is used for $\mathcal{L}\{f(t)\} = F(s)$

FRAME 9

In solving d.e.'s we shall need $F(s)$ for various forms of $f(t)$ and also $f(t)$ for various forms of $F(s)$. To facilitate this we need to extend the table started in FRAME 6, and must obtain the L.T. of other forms of $f(t)$.

Another useful form of $f(t)$ is t. Find $\mathcal{L}\{t\}$, using the definition.

****

- SOLUTION BY LAPLACE TRANSFORM METHODS

9A

$$\mathcal{L}\{t\} = \int_0^\infty e^{-st} t \, dt = \left[-\frac{1}{s} e^{-st} t\right]_0^\infty + \frac{1}{s} \int_0^\infty e^{-st} dt \quad \text{using integration by parts}$$

$$= \frac{1}{s^2}$$

<u>Note</u> that $\mathcal{L}^{-1}\{\frac{1}{s^2}\} = t$

FRAME 10

To extend this result to $f(t) = t^n$, n being a positive integer:

$$\mathcal{L}\{t^n\} = \int_0^\infty e^{-st} t^n \, dt$$

This integral is evaluated by using a reduction formula.

Let $I_n = \int_0^\infty e^{-st} t^n \, dt$

$$= \left[-\frac{1}{s} e^{-st} t^n\right]_0^\infty + \frac{n}{s} \int_0^\infty e^{-st} t^{n-1} \, dt$$

$$= \frac{n}{s} I_{n-1}$$

Use this reduction formula to obtain the final result.

****

10A

$I_n = \frac{n}{s} I_{n-1}$

$I_{n-1} = \frac{n-1}{s} I_{n-2}$

$\vdots$

$I_1 = \frac{1}{s} I_0$

$I_0 = \frac{1}{s}$ $\quad \left[\text{Note that } t^0 = 1 \text{ and } \mathcal{L}\{1\} = \frac{1}{s} \text{ from (5A.1).}\right]$

$\therefore \quad I_n = \frac{n(n-1) \ldots \ldots 1}{s^{n+1}} = \frac{n!}{s^{n+1}}$

$\therefore \quad \mathcal{L}\{t^n\} = \frac{n!}{s^{n+1}}$

FRAME 11

Some properties of the Laplace Transform

In this and some later frames, we shall develop some properties of the Laplace Transform, which can be used in finding transforms of more complicated functions, without going back to the definition itself.

Theorem I

If $\mathcal{L}\{f(t)\} = F(s)$ then $\mathcal{L}\{kf(t)\} = kF(s)$, where k is any constant.
This follows from the definition:
$$\mathcal{L}\{kf(t)\} = \int_0^\infty e^{-st} kf(t)\, dt = k\int_0^\infty e^{-st} f(t)\, dt = kF(s)$$

Theorem II

If $\mathcal{L}\{f(t)\} = F(s)$ and $\mathcal{L}\{g(t)\} = G(s)$ then
$$\mathcal{L}\{f(t) \pm g(t)\} = F(s) \pm G(s).$$

Use the definition to show this result.

****

11A

$$\mathcal{L}\{f(t) \pm g(t)\} = \int_0^\infty e^{-st}\{f(t) \pm g(t)\}dt$$
$$= \int_0^\infty e^{-st} f(t)\, dt \pm \int_0^\infty e^{-st} g(t)\, dt$$
$$= F(s) \pm G(s)$$

FRAME 12

We can now use these theorems to obtain the transforms of some more functions. For example, we know that $e^{i\omega t} = \cos \omega t + i \sin \omega t$,

hence $\mathcal{L}\{e^{i\omega t}\} = \mathcal{L}\{\cos \omega t\} + i\mathcal{L}\{\sin \omega t\}$ using Theorems I and II.

But $\mathcal{L}\{e^{i\omega t}\} = \dfrac{1}{s - i\omega}$ from (4A.1), ω being a constant

$= \dfrac{s + i\omega}{s^2 + \omega^2}$ on rationalisation.

$\therefore \mathcal{L}\{\cos \omega t\} + i\mathcal{L}\{\sin \omega t\} = \dfrac{s + i\omega}{s^2 + \omega^2}$

- SOLUTION BY LAPLACE TRANSFORM METHODS

FRAME 12 continued

Equating real and imaginary parts we get the two important results:

$$\mathcal{L}\{\cos \omega t\} = \frac{s}{s^2 + \omega^2}$$

$$\mathcal{L}\{\sin \omega t\} = \frac{\omega}{s^2 + \omega^2}$$

By using the identity $\sin(A + B) = \sin A \cos B + \cos A \sin B$, and the above results, find $\mathcal{L}\{\sin(\omega t + \phi)\}$ where ϕ is a constant.

****

12A

$$\mathcal{L}\{\sin(\omega t + \phi)\} = \mathcal{L}\{\sin \omega t \cos \phi + \cos \omega t \sin \phi\}$$

$$= \frac{\omega \cos \phi + s \sin \phi}{s^2 + \omega^2}$$

FRAME 13

Now, with the results so far obtained, find the Laplace Transforms of the following functions. These transform pairs will be added to the standard table.

(1) $\cos(\omega t + \phi)$

(2) $\sinh \beta t$
(3) $\cosh \beta t$ } where β is a constant. HINT: Express $\sinh \beta t$ and $\cosh \beta t$ in exponential form.

(4) $\frac{1}{a - b}(e^{at} - e^{bt})$, a and b constant

(5) $\frac{1}{a - b}(ae^{at} - be^{bt})$

(6) $1 - e^{\alpha t}$

(7) $1 - \cos \omega t$

****

(1) $\dfrac{s \cos \phi - \omega \sin \phi}{s^2 + \omega^2}$

(2) $\mathcal{L}\{\sinh \beta t\} = \mathcal{L}\{\tfrac{1}{2}(e^{\beta t} - e^{-\beta t})\}$

$= \dfrac{1}{2}\left[\dfrac{1}{s - \beta} - \dfrac{1}{s + \beta}\right]$

$= \dfrac{\beta}{s^2 - \beta^2}$

(3) $\dfrac{s}{s^2 - \beta^2}$, by similar method to (2)

(4) $\dfrac{1}{(s - a)(s - b)}$

(5) $\dfrac{s}{(s - a)(s - b)}$

(6) $-\dfrac{\alpha}{s(s - \alpha)}$

(7) $\dfrac{\omega^2}{s(s^2 + \omega^2)}$

NOTE: (4), (5), (6) and (7) are widely used in finding inverse Laplace Transforms. E.g. $\mathcal{L}^{-1}\left\{\dfrac{1}{(s - a)(s - b)}\right\} = \dfrac{1}{a - b}(e^{at} - e^{bt})$

FRAME 14

Let us now obtain another useful result.

Theorem III

If $\mathcal{L}\{f(t)\} = F(s)$, then $\mathcal{L}\{e^{\alpha t}f(t)\} = F(s - \alpha)$, where α is a constant.

This property follows from the definition:

$$\mathcal{L}\{e^{\alpha t}f(t)\} = \int_0^\infty e^{-st}e^{\alpha t}f(t)\,dt$$

$$= \int_0^\infty e^{-(s-\alpha)t}f(t)\,dt \qquad (14.1)$$

$$\text{Now } F(s) = \int_0^\infty e^{-st}f(t)\,dt \qquad (14.2)$$

- SOLUTION BY LAPLACE TRANSFORM METHODS

FRAME 14 continued

In (14.2) we have F(s), where s is the parameter in the integrand. You will see that the integral in (14.1) takes the same form with $(s - \alpha)$ as the parameter. We may therefore write

$$F(s - \alpha) = \int_0^\infty e^{-(s-\alpha)t} f(t)\, dt.$$

$$\therefore \quad \mathcal{L}\{e^{\alpha t} f(t)\} = F(s - \alpha)$$

NOTE: $F(s - \alpha)$ is obtained by replacing every s in F(s) by $(s - \alpha)$. This enables us to simply write down $\mathcal{L}\{e^{\alpha t} f(t)\}$ when $\mathcal{L}\{f(t)\}$ is known.

FRAME 15

To illustrate, let us find $\mathcal{L}\{e^{\alpha t} t\}$.

We already know that $\mathcal{L}\{t\} = \dfrac{1}{s^2}$ (see 9A)

$\therefore$ Using Theorem III, we have $\mathcal{L}\{e^{\alpha t} t\} = \dfrac{1}{(s - \alpha)^2}$, replacing s by $(s - \alpha)$.

Using the transforms of functions already obtained, write down the transforms of the following:

$e^{\alpha t} t^n$, $e^{\alpha t} \sin \omega t$, $e^{\alpha t} \cos \omega t$.

****

15A

We know that $\mathcal{L}\{t^n\} = \dfrac{n!}{s^{n+1}}$

$\therefore \quad \mathcal{L}\{e^{\alpha t} t^n\} = \dfrac{n!}{(s - \alpha)^{n+1}}$ *using Theorem III*

In a similar manner,

$$\mathcal{L}\{e^{\alpha t} \sin \omega t\} = \dfrac{\omega}{(s - \alpha)^2 + \omega^2}$$

$$\mathcal{L}\{e^{\alpha t} \cos \omega t\} = \dfrac{s - \alpha}{(s - \alpha)^2 + \omega^2}$$

FRAME 16

Use of the table of transforms

The transform pairs obtained so far are summarised in the form of a table, for your convenience, in the APPENDIX at the end of this programme. You will find that, in solving problems using L.T. methods, you will often need to refer to such a table, both for obtaining the transform of f(t) and the inverse transform of F(s).

E.g. $\quad \mathcal{L}\{t^2\} = \dfrac{2}{s^3}, \qquad \mathcal{L}^{-1}\{\dfrac{1}{s^3}\} = \dfrac{1}{2}t^2$

$$\mathcal{L}\{e^{5t} - e^{2t}\} = \dfrac{3}{(s-5)(s-2)}, \quad \text{while if} \quad \mathcal{L}^{-1}\{\dfrac{1}{s^2 - 7s + 10}\}$$

were required, we first factorise the denominator to give one of the standard forms in the table.

$$\mathcal{L}^{-1}\{\dfrac{1}{s^2 - 7s + 10}\} = \mathcal{L}^{-1}\{\dfrac{1}{(s-5)(s-2)}\}$$
$$= \dfrac{1}{5 - 2}(e^{5t} - e^{2t})$$
$$= \dfrac{1}{3}(e^{5t} - e^{2t})$$

Now try these examples.

1. Find the L.T. of the following functions of t:

   (i) $\quad t^2 - 2 + e^{-t} - \sin 3t$

   (ii) $\quad \sin(t + \pi/4) - 2 \cosh 2t$

   (iii) $\quad 3e^{\frac{1}{2}t} t^3$

   (iv) $\quad 5e^{-2t} \cos \pi t$

2. Find the inverse transforms of the following functions of s:

   (i) $\quad \dfrac{3}{2s - 1}$ (ii) $\quad \dfrac{s + 2}{s^2 + 9}$

   (iii) $\quad \dfrac{1}{s(s^2 + 4)}$ (iv) $\quad \dfrac{5}{(s + 2)^2 + 9}$

****

- SOLUTION BY LAPLACE TRANSFORM METHODS

16A

1. (i) $\dfrac{2}{s^3} - \dfrac{2}{s} + \dfrac{1}{s+1} - \dfrac{3}{s^2+9}$

   (ii) $\dfrac{s+1}{\sqrt{2}(s^2+1)} - \dfrac{2s}{s^2-4}$

   (iii) $\dfrac{18}{(s-\frac{1}{2})^4}$

   (iv) $\dfrac{5(s+2)}{(s+2)^2+\pi^2}$

2. (i) $\dfrac{3}{2}e^{\frac{1}{2}t}$ ($\dfrac{3}{2s-1}$ is written as $\dfrac{3}{2}\dfrac{1}{s-\frac{1}{2}}$)

   (ii) $\cos 3t + \dfrac{2}{3}\sin 3t$

   (iii) $\dfrac{1}{4}(1 - \cos 2t)$

   (iv) $\dfrac{5}{3}e^{-2t}\sin 3t$

FRAME 17

If you look at example 2 (iv) of the previous frame,

i.e. $\mathcal{L}^{-1}\{\dfrac{5}{(s+2)^2+9}\} = \dfrac{5}{3}e^{-2t}\sin 3t$

you will see that $F(s)$ is $\dfrac{5}{s^2+4s+13}$. Thus, if you were required to find $\mathcal{L}^{-1}\{\dfrac{5}{s^2+4s+13}\}$, you would first need to "complete the square" in the denominator before using the standard form in the table.

For practice, find $\mathcal{L}^{-1}\{\dfrac{1}{s^2-3s+3}\}$

****

17A

$$\mathcal{L}^{-1}\{\dfrac{1}{s^2-3s+3}\} = \mathcal{L}^{-1}\left\{\dfrac{1}{(s-\tfrac{3}{2})^2+(\tfrac{\sqrt{3}}{2})^2}\right\}$$

$$= \dfrac{2}{\sqrt{3}}e^{3t/2}\sin\dfrac{\sqrt{3}}{2}t.$$

FRAME 18

However, if $\mathcal{L}^{-1}\{\frac{s}{s^2 + 4s + 13}\}$ were required, where s appears in the numerator, we would proceed as follows:

$$\mathcal{L}^{-1}\{\frac{s}{s^2 + 4s + 13}\} = \mathcal{L}^{-1}\{\frac{s}{(s + 2)^2 + 9}\}$$

$$= \mathcal{L}^{-1}\{\frac{s + 2 - 2}{(s + 2)^2 + 9}\}$$

$$= \mathcal{L}^{-1}\{\frac{s + 2}{(s + 2)^2 + 9} - \frac{2}{3}\frac{3}{(s + 2)^2 + 9}\}$$

$$= e^{-2t} \cos 3t - \frac{2}{3} e^{-2t} \sin 3t$$

We write s as $s + 2 - 2$ to get the standard forms $\frac{s - \alpha}{(s - \alpha)^2 + \omega^2}$ and $\frac{\omega}{(s - \alpha)^2 + \omega^2}$, two of the results in the APPENDIX.

Use the above procedure to obtain $\mathcal{L}^{-1}\{\frac{3s - 4}{s^2 - 6s + 10}\}$

****

18A

$$\mathcal{L}^{-1}\{\frac{3s - 4}{(s - 3)^2 + 1}\} = \mathcal{L}^{-1}\{\frac{3(s - 3) + 5}{(s - 3)^2 + 1}\}$$

$$= 3e^{3t} \cos t + 5e^{3t} \sin t$$

FRAME 19

We will now obtain some more results for inclusion in the table.

In FRAME 15 we showed that $\mathcal{L}\{e^{\alpha t} t\} = \frac{1}{(s - \alpha)^2}$.

In this, put $\alpha = i\omega$ giving

$$\mathcal{L}\{te^{i\omega t}\} = \frac{1}{(s - i\omega)^2}$$

$$= \frac{(s + i\omega)^2}{(s^2 + \omega^2)^2} \quad \text{on rationalisation}$$

$$\therefore \mathcal{L}\{t(\cos \omega t + i \sin \omega t)\} = \frac{s^2 - \omega^2 + 2i\omega s}{(s^2 + \omega^2)^2}$$

SOLUTION BY LAPLACE TRANSFORM METHODS

FRAME 19 continued

On separating real and imaginary parts we get the two useful results:

$$\mathcal{L}\{t \cos \omega t\} = \frac{s^2 - \omega^2}{(s^2 + \omega^2)^2}$$

$$\mathcal{L}\{t \sin \omega t\} = \frac{2\omega s}{(s^2 + \omega^2)^2}$$

Now write down $\mathcal{L}\{\sin \omega t - \omega t \cos \omega t\}$

****

19A

$$\mathcal{L}\{sin\ \omega t\ -\ \omega t\ cos\ \omega t\} = \frac{\omega}{s^2 + \omega^2} - \frac{\omega(s^2 - \omega^2)}{(s^2 + \omega^2)^2}$$

$$= \frac{2\omega^3}{(s^2 + \omega^2)^2}$$

This gives the useful result $\mathcal{L}^{-1}\{\frac{2\omega^3}{(s^2 + \omega^2)^2}\} = sin\ \omega t\ -\ \omega t\ cos\ \omega t$

FRAME 20

Use of partial fractions in finding inverse transforms

The three results in the previous frame are sometimes useful in finding the inverse transforms of certain rational functions when they are split up into partial fractions. The use of partial fractions in finding inverse transforms is illustrated by the examples which follow.

Example 1   Find $\mathcal{L}^{-1}\{\frac{1}{(s - a)(s - b)}\}$

The rational function $\frac{1}{(s - a)(s - b)}$, which has a denominator that factorises, can be resolved into partial fractions,

i.e. $\frac{1}{(s - a)(s - b)} = \frac{A}{s - a} + \frac{B}{s - b}$

giving $A = \frac{1}{a - b}$ and $B = -\frac{1}{a - b}$.

Hence $\mathcal{L}^{-1}\{\frac{1}{(s - a)(s - b)}\} = \frac{1}{a - b} \mathcal{L}^{-1}\{\frac{1}{s - a} - \frac{1}{s - b}\}$

$$= \frac{1}{a - b}(e^{at} - e^{bt})$$

F

FRAME 20 continued

In FRAME 13 Example (4) you obtained this result as

$$\mathcal{L}\{\frac{1}{a-b}(e^{at} - e^{bt})\} = \frac{1}{(s-a)(s-b)}$$

The alternative method given in this frame shows the use of partial fractions in finding inverse transforms.

FRAME 21

Example 2. Find $\mathcal{L}^{-1}\{\frac{2s+3}{(s-1)(s-2)(s-3)}\}$

This example can only be done by using partial fractions. Resolve the function into partial fractions and hence find the inverse $f(t)$.

****

21A

$$\frac{2s+3}{(s-1)(s-2)(s-3)} = \frac{5/2}{s-1} - \frac{7}{s-2} + \frac{9/2}{s-3}$$

Hence $f(t) = \frac{5}{2}e^t - 7e^{2t} + \frac{9}{2}e^{3t}$

FRAME 22

Example 3 Find $\mathcal{L}^{-1}\{\frac{8s+10}{(s+2)^2(s^2+2s+2)}\}$

****

22A

$$\frac{8s+10}{(s+2)^2(s^2+2s+2)} = \frac{A}{s+2} + \frac{B}{(s+2)^2} + \frac{Cs+D}{s^2+2s+2}$$

giving $A = 1$, $B = -3$, $C = -1$ and $D = 3$.

$$\mathcal{L}^{-1}\{\frac{8s+10}{(s+2)^2(s^2+2s+2)}\} = \mathcal{L}^{-1}\{\frac{1}{s+2} - \frac{3}{(s+2)^2} - \frac{s+1-4}{(s+1)^2+1}\}$$

$$= e^{-2t} - 3te^{-2t} - e^{-t}\cos t + 4e^{-t}\sin t$$

- SOLUTION BY LAPLACE TRANSFORM METHODS

FRAME 23

Example 4   Find $\mathcal{L}^{-1}\{\frac{4s^2 + 3s + 16}{s(s^2 + 4)^2}\}$

****

23A

$$\frac{4s^2 + 3s + 16}{s(s^2 + 4)^2} = \frac{A}{s} + \frac{Bs + C}{s^2 + 4} + \frac{Ds + E}{(s^2 + 4)^2}$$

giving $A = 1$, $\quad B = -1$, $\quad C = 0$, $\quad D = 0$ *and* $E = 3$.

$$\mathcal{L}^{-1}\{\frac{4s^2 + 3s + 16}{s(s^2 + 4)^2}\} = \mathcal{L}^{-1}\{\frac{1}{s} - \frac{s}{s^2 + 4} + \frac{3}{(s^2 + 4)^2}\}$$

$$= 1 - \cos 2t + \frac{3}{16}(\sin 2t - 2t \cos 2t)$$

FRAME 24

Transforms of derivatives

In solving differential equations by Laplace Transform methods, the transforms of differential coefficients will be required. If the d.e. is

$$a\frac{d^2x}{dt^2} + b\frac{dx}{dt} + cx = f(t),$$

the transforms of x, $\frac{dx}{dt}$ and $\frac{d^2x}{dt^2}$ are needed.

$$\mathcal{L}\{x\} = \int_0^\infty e^{-st} x \, dt = \bar{x}, \quad \text{say}.$$

The bar is an alternative notation for the Laplace Transform e.g. $\mathcal{L}\{q\}$ could be written as $\bar{q}$.

$$\mathcal{L}\{\frac{dx}{dt}\} = \int_0^\infty e^{-st} \frac{dx}{dt} dt$$

$$= \left[e^{-st} x\right]_0^\infty + s\int_0^\infty e^{-st} x \, dt \quad \text{on integration by parts}$$

$$= 0 - x_0 + s\bar{x} \quad \text{where } x_0 \text{ is the value of } x \text{ at } t = 0$$

$$= s\bar{x} - x_0$$

Now find $\mathcal{L}\{\frac{d^2x}{dt^2}\}$ using a similar procedure, assuming $\frac{dx}{dt} = x_1$ at $t = 0$.

****

24A

$$\mathcal{L}\{\tfrac{d^2x}{dt^2}\} = \int_0^\infty e^{-st}\tfrac{d^2x}{dt^2}\,dt$$

$$= \left[e^{-st}\tfrac{dx}{dt}\right]_0^\infty + s\int_0^\infty e^{-st}\tfrac{dx}{dt}\,dt$$

$$= 0 - \left(\tfrac{dx}{dt}\right)_{t=0} + s(s\bar{x} - x_0)$$

$$= s^2\bar{x} - sx_0 - x_1 \quad \text{where} \quad x_1 = \left(\tfrac{dx}{dt}\right)_{t=0}$$

FRAME 25

So far, we have

$$\mathcal{L}\{x\} = \bar{x}$$

$$\mathcal{L}\{\tfrac{dx}{dt}\} = s\bar{x} - x_0$$

$$\mathcal{L}\{\tfrac{d^2x}{dt^2}\} = s^2\bar{x} - sx_0 - x_1$$

Similarly, $\mathcal{L}\{\tfrac{d^3x}{dt^3}\} = s^3\bar{x} - s^2x_0 - sx_1 - x_2$,

$\mathcal{L}\{\tfrac{d^4x}{dt^4}\} = s^4\bar{x} - s^3x_0 - s^2x_1 - sx_2 - x_3$,

and in general,

$$\mathcal{L}\{\tfrac{d^nx}{dt^n}\} = s^n\bar{x} - s^{n-1}x_0 - s^{n-2}x_1 - \ldots - x_{n-1}$$

where x_{n-1} is the value of $\tfrac{d^{n-1}x}{dt^{n-1}}$ at $t = 0$.

FRAME 26

Solution of differential equations

We have now completed the table of Laplace Transform pairs in the APPENDIX. For the purposes of this programme, this table is sufficiently comprehensive. The stage has now been reached when we can begin solving differential equations with constant coefficients by Laplace Transform methods.

- SOLUTION BY LAPLACE TRANSFORM METHODS

FRAME 26 continued

Consider, for example, the solution of the d.e.

$$\frac{d^2x}{dt^2} - 3\frac{dx}{dt} + 2x = 2e^{-4t} \qquad (26.1)$$

given that $x = 0$, $\frac{dx}{dt} = 1$ at $t = 0$ (i.e. $x_o = 0$, $x_1 = 1$).

Transforming the L.H.S. of (26.1) we get

$$(s^2\bar{x} - sx_o - x_1) - 3(s\bar{x} - x_o) + 2\bar{x}$$

Hence the transformed equation is

$$(s^2\bar{x} - 1) - 3s\bar{x} + 2\bar{x} = \frac{2}{s+4}$$

i.e. $\quad (s^2 - 3s + 2)\bar{x} = \frac{2}{s+4} + 1 = \frac{s+6}{s+4}$

In FRAMES 2 and 3 it was pointed out that by using the Laplace transformation a d.e. is transformed to an algebraic equation incorporating the given initial conditions. You will notice that this is what has happened here, where we now have an equation in $\bar{x}$, the transform of the required solution.

This equation gives $\quad \bar{x} = \dfrac{s+6}{(s^2 - 3s + 2)(s+4)} = \dfrac{s+6}{(s-1)(s-2)(s+4)}$

The required particular solution x can be obtained by finding the inverse transform of $\dfrac{s+6}{(s-1)(s-2)(s+4)}$. You have already seen how to use partial fractions to do this, so now find x.

****

26A

$$x = \mathcal{L}^{-1}\{\frac{s+6}{(s-1)(s-2)(s+4)}\}$$

$$= \mathcal{L}^{-1}\{\frac{A}{s-1} + \frac{B}{s-2} + \frac{C}{s+4}\} \quad \text{where } A = -\frac{7}{5}, \ B = \frac{4}{3}, \ C = \frac{1}{15}$$

$$= -\frac{7}{5}e^t + \frac{4}{3}e^{2t} + \frac{1}{15}e^{-4t}$$

FRAME 27

When using either trial solution or D-operator methods you will have met "cases of failure" which had to be dealt with in a special way. An example is

$$\frac{d^2x}{dt^2} - \frac{dx}{dt} - 2x = 5e^{2t}$$

where the C.F. $(Ae^{2t} + Be^{-t})$ contains the function e^{2t} appearing on the R.H.S.

Another advantage of Laplace Transform methods is that, when they are used there are no cases of failure. You can see this for yourself when solving the above d.e. with the conditions $x = 0$, $\frac{dx}{dt} = 0$ at $t = 0$.

First write down the transformed equation and find $\bar{x}$ as a function of s.

****

27A

$$s^2\bar{x} - s\bar{x} - 2\bar{x} = \frac{5}{s-2}$$

$$(s^2 - s - 2)\bar{x} = \frac{5}{s-2}$$

$$\bar{x} = \frac{5}{(s^2 - s - 2)(s - 2)}$$

$$= \frac{5}{(s+1)(s-2)^2}$$

FRAME 28

You now have $\bar{x} = \dfrac{5}{(s+1)(s-2)^2}$

By resolving the R.H.S. into partial fractions, find its inverse transform i.e. x.

****

28A

$$\bar{x} = \frac{A}{s+1} + \frac{B}{s-2} + \frac{C}{(s-2)^2} \quad \text{where } A = \frac{5}{9}, \quad B = -\frac{5}{9}, \quad C = \frac{5}{3}.$$

$$x = \frac{5}{9}e^{-t} - \frac{5}{9}e^{2t} + \frac{5}{3}te^{2t}$$

You will notice that there is no breakdown in the usual procedure.

- SOLUTION BY LAPLACE TRANSFORM METHODS 4:19

FRAME 29

The method discussed in FRAME 26 is used to solve d.e.'s with functions other than exponential ones on the R.H.S.

For example, let us consider the solution of
$$2\frac{d^2x}{dt^2} + 5\frac{dx}{dt} - 3x = t - 4$$
with the conditions $x = 0$ and $\frac{dx}{dt} = 2$ at $t = 0$.

The transformation of the L.H.S. will be dealt with as before.
The transform of the R.H.S. is looked up in the table, from which it is found that $\mathcal{L}\{t\} = \frac{1}{s^2}$ and $\mathcal{L}\{4\} = \frac{4}{s}$.

Proceeding in this way, show that $\bar{x} = \frac{2s - 1}{s^2(s + 3)}$ and hence complete the solution.

****

29A

$$2(s^2\bar{x} - 2) + 5s\bar{x} - 3\bar{x} = \frac{1}{s^2} - \frac{4}{s}$$

$$(2s^2 + 5s - 3)\bar{x} = \frac{1}{s^2} - \frac{4}{s} + 4 = \frac{1 - 4s + 4s^2}{s^2}$$

$$\bar{x} = \frac{(2s - 1)^2}{s^2(2s - 1)(s + 3)} = \frac{2s - 1}{s^2(s + 3)}$$

$$= \frac{A}{s} + \frac{B}{s^2} + \frac{C}{s + 3} \quad \text{where} \quad A = \frac{7}{9}, \quad B = -\frac{1}{3}, \quad C = -\frac{7}{9}$$

$$\therefore \quad x = \frac{7}{9} - \frac{1}{3}t - \frac{7}{9}e^{-3t}$$

FRAME 30

Another example: Solve the d.e. $\frac{d^2y}{dt^2} + y = \cos t$ given that $y = 0$ and $\frac{dy}{dt} = -1$ at $t = 0$.

As before, transform the d.e. to obtain $\bar{y}$.

****

30A

$$\{s^2 \bar{y} - (-1)\} + \bar{y} = \frac{s}{s^2 + 1}$$

$$(s^2 + 1)\bar{y} = \frac{s}{s^2 + 1} - 1$$

$$\bar{y} = \frac{s}{(s^2 + 1)^2} - \frac{1}{s^2 + 1}$$

You will notice that we have not put the R.H.S. over a common denominator, because, as it stands, it consists of standard forms given in the table.

FRAME 31

Now use the table to find y.

****

31A

$$y = \tfrac{1}{2} t \sin t - \sin t$$
$$= \tfrac{1}{2}(t - 2) \sin t$$

FRAME 32

Let us now solve the d.e.

$$\frac{d^2 y}{dx^2} + \frac{dy}{dx} - 2y = 5e^{-x} \sin 2x \qquad (32.1)$$

with the conditions $y = 1$, $\frac{dy}{dx} = 0$ at $x = 0$.

The independent variable here is x, instead of t. Obviously, you will have to substitute x for t in the table.

Transform (32.1) to obtain $\bar{y}$.

****

32A

$$(s^2 \bar{y} - s) + (s\bar{y} - 1) - 2\bar{y} = 5 \frac{2}{(s + 1)^2 + 4}$$

$$\text{giving} \quad \bar{y} = \frac{s^3 + 3s^2 + 7s + 15}{(s - 1)(s + 2)(s^2 + 2s + 5)}$$

- SOLUTION BY LAPLACE TRANSFORM METHODS 4:21

FRAME 33

Resolve $\bar{y}$ into partial fractions to obtain

$$\bar{y} = \frac{13/12}{s-1} + \frac{-1/3}{s+2} + \frac{s/4 - 5/4}{s^2 + 2s + 5}.$$

Then find y. If you have any difficulty in inverting the third fraction, look back to answer frame 22A.

****

33A

$$\frac{s/4 - 5/4}{s^2 + 2s + 5} = \frac{1}{4}\left\{\frac{s+1}{(s+1)^2 + 4} - \frac{6}{(s+1)^2 + 4}\right\}$$

hence $y = \frac{13}{12} e^x - \frac{1}{3} e^{-2x} + \frac{1}{4}\left(e^{-x} \cos 2x - 3e^{-x} \sin 2x\right)$

$ = \frac{13}{12} e^x - \frac{1}{3} e^{-2x} + \frac{1}{4} e^{-x}(\cos 2x - 3 \sin 2x)$

FRAME 34

So far we have only dealt with second order d.e.'s. We shall now consider a higher order d.e. to illustrate the general applicability of the method.

Find the solution of $\frac{d^4 y}{dt^4} - k^4 y = 0$, where k is a constant, given that

$y = 1$, $\frac{dy}{dt} = \frac{d^2 y}{dt^2} = \frac{d^3 y}{dt^3} = 0$ at $t = 0$.

From the formula in the table, $\mathcal{L}\{\frac{d^4 y}{dt^4}\} = s^4 \bar{y} - s^3 y_0 - s^2 y_1 - s y_2 - y_3$

giving the transformed equation as

$$(s^4 \bar{y} - s^3) - k^4 \bar{y} = 0$$

i.e. $\bar{y} = \frac{s^3}{s^4 - k^4}.$

Now resolve $\bar{y}$ into partial fractions and complete the solution.

****

4:22   DIFFERENTIAL EQUATIONS WITH CONSTANT COEFFICIENTS

34A

$$\bar{y} = \frac{A}{s-k} + \frac{B}{s+k} + \frac{Cs+D}{s^2+k^2}$$

giving $A = \frac{1}{4}$, $B = \frac{1}{4}$, $C = \frac{1}{2}$ and $D = 0$.

Hence $y = \frac{1}{4}e^{kt} + \frac{1}{4}e^{-kt} + \frac{1}{2}\cos kt$

$$= \frac{1}{2}(\cosh kt + \cos kt)$$

FRAME 35

We shall now give two examples of the use of Laplace Transform methods in the solution of d.e.'s arising in physical problems.

Application 1

A battery of constant voltage E_o is connected to an inductance L H in series with a capacitance C F. If the initial charge and current are both zero, show that when the current reverses (next becomes zero) the condenser is charged to a voltage $2E_o$.

If i A is the current flowing in the circuit and q C is the charge on the condenser plates at time t, then on equating the total potential drop across each of the components to E_o we have the d.e.

$$L\frac{di}{dt} + \frac{q}{C} = E_o.$$

Noting that $i = \frac{dq}{dt}$ this becomes

$$L\frac{d^2q}{dt^2} + \frac{q}{C} = E_o$$

or $\frac{d^2q}{dt^2} + \omega^2 q = \frac{E_o}{L}$ where $\omega^2 = \frac{1}{LC}$

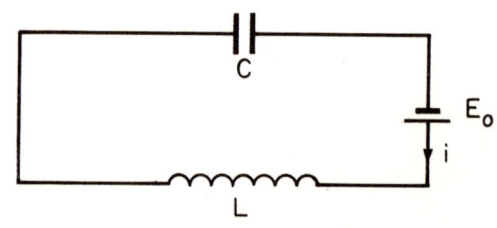

This is the differential equation for the charge q at any time t.

Use the Laplace Transform method to obtain q.

****

SOLUTION BY LAPLACE TRANSFORM METHODS

35A

$$(s^2 + \omega^2)\bar{q} = \frac{E_o}{Ls} \text{ using the initial conditions } i\left(=\frac{dq}{dt}\right) = 0, \; q = 0 \text{ at } t = 0.$$

$$\bar{q} = \frac{E_o}{Ls(s^2 + \omega^2)}$$

This is a standard form given in the table.

$$\therefore \; q = \frac{E_o}{L\omega^2}(1 - \cos \omega t)$$

$$= E_o C(1 - \cos \omega t)$$

FRAME 36

Having obtained q, find i by differentiation and find when the current is next zero (after t = 0). Hence complete the solution of the problem.

****

36A

$$q = E_o C(1 - \cos \omega t)$$

$$i = \frac{dq}{dt} = \omega E_o C \sin \omega t$$

The current is next zero (after t = 0) when $\omega t = \pi$ and at this time the charge $q = 2E_o C$.

∴ Condenser is charged to voltage $\frac{q}{C} = 2E_o$.

FRAME 37

Application 2

As a second example we consider the problem of a light rod clamped horizontally at one end (x = 0), freely hinged at the other (x = ℓ) and subjected to a horizontal thrust P at the clamped end.

FRAME 37 continued

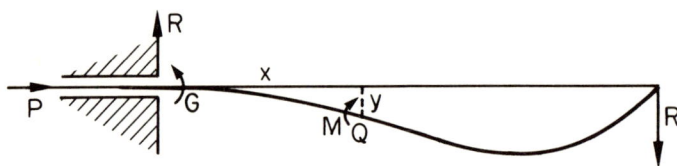

Taking moments about any point Q, on the rod, for forces to the left of Q we obtain
$$M - G + Py + Rx = 0 \qquad (37.1)$$
where $M(= EIy'')$ is the bending moment, G is the clamping couple and R is the reaction at the ends. EI is the constant flexural rigidity of the beam.
Taking moments about the end $x = 0$ gives $G = R\ell$ and (37.1) becomes
$$EIy'' + Py = R(\ell - x).$$
We shall now solve this to obtain the deflection y at position x.
Writing $P/EI = n^2$ and $R/EI = k$, the equation becomes
$$y'' + n^2 y = k(\ell - x).$$
First apply the boundary conditions $y = y' = 0$ at the end $x = 0$ to obtain $\bar{y}$ in the form
$$\bar{y} = k \{ \frac{\ell}{s(s^2 + n^2)} - \frac{1}{s^2(s^2 + n^2)} \}$$
(The first fraction is to be found in the table and its inverse can be written down at once. This is the reason why we have preferred not to combine it with the remaining term on the R.H.S.)

Then express $\dfrac{1}{s^2(s^2 + n^2)}$ in partial fractions.

****

37A

$$s^2 \bar{y} + n^2 \bar{y} = k(\frac{\ell}{s} - \frac{1}{s^2})$$
$$\bar{y} = k \{ \frac{\ell}{s(s^2 + n^2)} - \frac{1}{s^2(s^2 + n^2)} \}$$

$\dfrac{1}{s^2(s^2 + n^2)}$ *is a function of* s^2 *and so the partial fractions can be taken as* $\dfrac{A}{s^2} + \dfrac{B}{s^2 + n^2}$. *(If in doubt about this, put* $s^2 = t$.*)*

- SOLUTION BY LAPLACE TRANSFORM METHODS 4:25

37A continued

$$A = \frac{1}{n^2} \qquad B = \frac{-1}{n^2}$$

$$\frac{1}{s^2(s^2 + n^2)} = \frac{1}{n^2}\left(\frac{1}{s^2} - \frac{1}{s^2 + n^2}\right)$$

FRAME 38

We now have

$$\bar{y} = k\left\{\frac{\ell}{s(s^2 + n^2)} - \frac{1}{n^2}\left(\frac{1}{s^2} - \frac{1}{s^2 + n^2}\right)\right\}$$

Use the table in the APPENDIX to find y.

****

38A

$$y = k\left\{\frac{\ell}{n^2}(1 - \cos nx) - \frac{x}{n^2} + \frac{1}{n^3} \sin nx\right\}$$

$$= \frac{R}{P}\left\{\ell\left(1 - \cos\sqrt{\frac{P}{EI}}\, x\right) - x + \sqrt{\frac{EI}{P}} \sin\sqrt{\frac{P}{EI}}\, x\right\}$$

FRAME 39

Miscellaneous Examples

In this frame a collection of miscellaneous examples is given for you to try. Answers are supplied in FRAME 40 and hints have been provided in some cases.

1. Using the table in the APPENDIX, find the Laplace transforms of the following functions of t:

   (i) $(\ell - t)^2$

   (ii) $e^{\alpha t} \cosh \beta t$

   (iii) $\frac{t}{2} e^{-t} \sin 3t$.

2. Find the inverse transforms of the following:

   (i) $\dfrac{1}{(s^2 + n^2)(s^2 + \omega^2)}$

   (ii) $\dfrac{s + 3}{s^2 + 4s + 1}$

   (iii) $\dfrac{4s + 2}{(s^2 + 6s + 13)^2}$

FRAME 39 continued

3. Solve the following differential equations with the given initial conditions:

(i) $\quad 2\dfrac{d^2y}{dx^2} - 3\dfrac{dy}{dx} - 5y = 0 \qquad\qquad y = 9,\ \dfrac{dy}{dx} = -2 \text{ at } x = 0$

(ii) $\quad \dfrac{d^2y}{dx^2} - 4\dfrac{dy}{dx} + 4y = 6xe^{2x} \qquad y = \dfrac{dy}{dx} = 1 \text{ at } x = 0$

(iii) $\quad \dfrac{d^2q}{dt^2} + 8\dfrac{dq}{dt} + 25q = 50 \qquad\qquad q = \dfrac{dq}{dt} = 0 \text{ at } t = 0$

(iv) $\quad \dfrac{d^2y}{dt^2} + 2\dfrac{dy}{dt} + y = \sin t \qquad\qquad y = 3,\ \dfrac{dy}{dt} = 1 \text{ at } t = 0$

(v) $\quad \dfrac{d^2y}{dx^2} - 3\dfrac{dy}{dx} + 2y = \dfrac{1}{2}e^x \qquad\qquad y = \dfrac{dy}{dx} = 0 \text{ at } x = 0$

(vi) $\quad 4\dfrac{d^2x}{dt^2} + 9x = \sin 2t \qquad\qquad x = 0,\ \dfrac{dx}{dt} = \dfrac{1}{2} \text{ at } t = 0$

(vii) $\quad \dfrac{d^2y}{dx^2} + 4\dfrac{dy}{dx} + 5y = 3e^{-2x} \qquad y = 4,\ \dfrac{dy}{dx} = -7 \text{ at } x = 0$

(viii) $\quad \dfrac{d^2x}{dt^2} + 6\dfrac{dx}{dt} + 10x = 50t \qquad\qquad x = 0,\ \dfrac{dx}{dt} = 1 \text{ at } t = 0$

(ix) $\quad \dfrac{d^2y}{dx^2} + 5\dfrac{dy}{dx} + 6y = 10e^{-2x}\sin 2x \qquad y = 0,\ \dfrac{dy}{dx} = -3 \text{ at } x = 0$

(x) $\quad \dfrac{d^2y}{dt^2} + 4\dfrac{dy}{dt} + 5y = 40\cos 5t \qquad\qquad y = 0,\ \dfrac{dy}{dt} = 1 \text{ at } t = 0$

4. When a resistance $R\ \Omega$ and an inductance L H are connected in series to a generator of voltage $E(t)$, the current i A at time t s after closing the circuit is given by the d.e.

$$L\dfrac{di}{dt} + Ri = E(t)$$

with the condition $i = 0$ at $t = 0$.

Find i in terms of t for each of the cases:
(i) $\quad E(t) = E_o$
(ii) $\quad E(t) = E_o \sin \omega t$

where E_o is a constant.

- SOLUTION BY LAPLACE TRANSFORM METHODS 4:27

FRAME 39 continued

5. In an oscillator circuit the d.e. for the charge q on the condenser at time t is $L\ddot{q} + \dfrac{q}{C} = E_o \cos pt$ where L, C, E_o and p are constants. Find q and the current i ($= \dot{q}$) at time t, if both q and i are initially zero, in the cases:
   (i) $p \neq \omega$
   (ii) $p = \omega$
   where $\omega^2 = 1/LC$.

6. When a beam of length ℓ is supported at its ends and carries a uniform load w per unit length, the deflection y at a distance x from one end satisfies the d.e.
$$\frac{d^4 y}{dx^4} = \frac{w}{EI}$$
and $y = \dfrac{d^2 y}{dx^2} = 0$ at each end i.e. at $x = 0$ and $x = \ell$.

Find the deflection at any point of the beam.

(HINT: Two of the given boundary conditions in this problem are not initial conditions, but apply at $x = \ell$, and you may wonder how to deal with this situation. The values of $\dfrac{dy}{dx}$ and $\dfrac{d^3 y}{dx^3}$ at $x = 0$ are not given – call them y_1 and y_3 respectively, and solve for y in the usual way. Then substitute the conditions at $x = \ell$ in your solution to find the unknowns y_1 and y_3.)

7. A body of unit mass moves along the x-axis under the action of an attractive force of magnitude $\omega^2 x$ directed towards the origin O. It is subject also to a frictional resistance of magnitude $2k\dot{x}$ and to a driving force $e^{-kt} \cos pt$. The equation of motion may, therefore, be written as
$$\ddot{x} + 2k\dot{x} + \omega^2 x = e^{-kt} \cos pt.$$
If the resistance is sufficiently small, so that $k < \omega$, we may write $\omega^2 - k^2 = n^2$, where n is real. Obtain the solution of the equation when
   (i) $p \neq n$
   (ii) $p = n$
assuming in both cases that the body starts from rest at the origin.

Answers to Miscellaneous Examples

1. (i) $\dfrac{\ell^2}{s} - \dfrac{2\ell}{s^2} + \dfrac{2}{s^3}$

   (ii) $\mathcal{L}\{\cosh \beta t\} = \dfrac{s}{s^2 - \beta^2}$

   $\therefore \mathcal{L}\{e^{\alpha t}\cosh \beta t\} = \dfrac{s - \alpha}{(s - \alpha)^2 - \beta^2}$

   (iii) $\mathcal{L}\{t \sin 3t\} = \dfrac{6s}{(s^2 + 9)^2}$

   $\therefore \mathcal{L}\{\tfrac{1}{2}e^{-t}t \sin 3t\} = \dfrac{1}{2}\dfrac{6(s + 1)}{\{(s + 1)^2 + 9\}^2}$

   $\qquad\qquad\qquad\quad = \dfrac{3(s + 1)}{(s^2 + 2s + 10)^2}$

2. (i) $\dfrac{1}{(s^2 + n^2)(s^2 + \omega^2)} = \dfrac{A}{s^2 + n^2} + \dfrac{B}{s^2 + \omega^2}$

   (If you thought it necessary to start with $\dfrac{As + B}{s^2 + n^2} + \dfrac{Cs + D}{s^2 + \omega^2}$ refer back to FRAME 37A.)

   $A = \dfrac{1}{\omega^2 - n^2}, \quad B = -\dfrac{1}{\omega^2 - n^2}$

   $\therefore \mathcal{L}^{-1}\{\dfrac{1}{(s^2 + n^2)(s^2 + \omega^2)}\} = \dfrac{1}{\omega^2 - n^2}(\dfrac{1}{n} \sin nt - \dfrac{1}{\omega} \sin \omega t)$

   (ii) $\mathcal{L}^{-1}\{\dfrac{s + 3}{s^2 + 4s + 1}\} = \mathcal{L}^{-1}\{\dfrac{s + 2}{(s + 2)^2 - 3} + \dfrac{1}{(s + 2)^2 - 3}\}$

   $\qquad\qquad\qquad\qquad = e^{-2t}(\cosh \sqrt{3}t + \dfrac{1}{\sqrt{3}} \sinh \sqrt{3}t)$

   (iii) $\mathcal{L}^{-1}\{\dfrac{4s + 2}{(s^2 + 6s + 13)^2}\} = \mathcal{L}^{-1}\{\dfrac{4(s + 3)}{[(s + 3)^2 + 4]^2} - \dfrac{10}{[(s + 3)^2 + 4]^2}\}$

   $\qquad\qquad\qquad\qquad = e^{-3t}t \sin 2t - \dfrac{5}{8}e^{-3t}(\sin 2t - 2t \cos 2t)$

- SOLUTION BY LAPLACE TRANSFORM METHODS 4:29

FRAME 40 continued

3. (i) Transformed equation is

$$2(s^2\bar{y} - 9s + 2) - 3(s\bar{y} - 9) - 5\bar{y} = 0$$

giving $\bar{y} = \dfrac{18s - 31}{2s^2 - 3s - 5} = \dfrac{4}{2s - 5} + \dfrac{7}{s + 1}$

$$y = 2e^{5x/2} + 7e^{-x}$$

(ii) Transformed equation: $(s^2\bar{y} - s - 1) - 4(s\bar{y} - 1) + 4\bar{y} = \dfrac{6}{(s - 2)^2}$

giving $\bar{y} = \dfrac{s - 3}{(s - 2)^2} + \dfrac{6}{(s - 2)^4}$

$= \dfrac{1}{s - 2} - \dfrac{1}{(s - 2)^2} + \dfrac{6}{(s - 2)^4}$

$$y = e^{2x} - xe^{2x} + x^3 e^{2x}$$

(iii) $\bar{q} = \dfrac{50}{s(s^2 + 8s + 25)}$

$= \dfrac{2}{s} - \dfrac{2(s + 4) + 8}{(s + 4)^2 + 9}$

$$q = 2 - 2e^{-4t} \cos 3t - \dfrac{8}{3} e^{-4t} \sin 3t$$

(iv) $y = \dfrac{7}{2} e^{-t} + \dfrac{9}{2} e^{-t} t - \dfrac{1}{2} \cos t$

(v) $y = \dfrac{1}{2}(e^{2x} - e^x - e^x x)$

(vi) $\bar{x} = \dfrac{2}{4s^2 + 9} + \dfrac{2}{(4s^2 + 9)(s^2 + 4)}$

$= \dfrac{2}{4s^2 + 9} + \dfrac{8/7}{4s^2 + 9} - \dfrac{2/7}{s^2 + 4}$

$= \dfrac{22/7}{4(s^2 + 9/4)} - \dfrac{2/7}{s^2 + 4}$

$x = \dfrac{11}{21} \sin \dfrac{3t}{2} - \dfrac{1}{7} \sin 2t$

3. (vii) $(s^2 + 4s + 5)\bar{y} = 4s + 9 + \dfrac{3}{s + 2}$

$\bar{y} = \dfrac{4(s + 2)}{(s + 2)^2 + 1} + \dfrac{1}{(s + 2)^2 + 1} + \dfrac{3}{(s + 2)\{(s + 2)^2 + 1\}}$

$y = 4e^{-2x}\cos x + e^{-2x}\sin x + 3e^{-2x}(1 - \cos x)$

$= e^{-2x}(\cos x + \sin x + 3)$

(viii) $x = 5t - 3 + e^{-3t}(3 \cos t + 5 \sin t)$

(ix) $\bar{y} = -\dfrac{3}{(s + 2)(s + 3)} + \dfrac{20}{(s + 2)(s + 3)(s^2 + 4s + 8)}$

$= -\dfrac{3}{(s + 2)(s + 3)} + \dfrac{5}{s + 2} - \dfrac{4}{s + 3} - \dfrac{s + 2}{(s + 2)^2 + 4}$

$\quad - \dfrac{4}{(s + 2)^2 + 4}$

$y = 3(e^{-3x} - e^{-2x}) + 5e^{-2x} - 4e^{-3x} - e^{-2x}\cos 2x - 2e^{-2x}\sin 2x$

$= e^{-2x}(2 - \cos 2x - 2 \sin 2x) - e^{-3x}$

(x) $y = \sin 5t - \cos 5t + e^{-2t}(\cos t - 2 \sin t)$

4. (i) $\bar{i} = \dfrac{E_o}{R}\dfrac{R/L}{s(s + R/L)}$

$i = \dfrac{E_o}{R}(1 - e^{-Rt/L})$

(ii) $i = \dfrac{E_o}{R^2 + \omega^2 L^2}(\omega L e^{-Rt/L} + R \sin \omega t - \omega L \cos \omega t)$

5. (i) $\bar{q} = \dfrac{E_o}{L}\dfrac{s}{(s^2 + p^2)(s^2 + \omega^2)}$

$= \dfrac{E_o}{L}\dfrac{1}{\omega^2 - p^2}\left\{\dfrac{s}{s^2 + p^2} - \dfrac{s}{s^2 + \omega^2}\right\}$

$q = \dfrac{E_o}{L(\omega^2 - p^2)}(\cos pt - \cos \omega t)$

$i = \dfrac{dq}{dt} = \dfrac{E_o}{L(\omega^2 - p^2)}(\omega \sin \omega t - p \sin pt)$

— SOLUTION BY LAPLACE TRANSFORM METHODS

FRAME 40 continued

5. (ii) $\quad q = \dfrac{E_o}{2\omega L} t \sin \omega t$

$\quad i = \dfrac{dq}{dt} = \dfrac{E_o}{2\omega L}(\sin \omega t + \omega t \cos \omega t)$

6. $\quad s^4 \bar{y} - s^2 y_1 - y_3 = \dfrac{w}{EIs}$

$\quad \bar{y} = \dfrac{y_1}{s^2} + \dfrac{y_3}{s^4} + \dfrac{w}{EIs^5}$

$\quad y = y_1 x + y_3 \dfrac{x^3}{6} + \dfrac{wx^4}{24EI}$

$\quad \dfrac{d^2 y}{dx^2} = y_3 x + \dfrac{wx^2}{2EI}$

Now substitute the conditions at $x = \ell$:

$y = 0$ at $x = \ell$ gives $0 = y_1 + y_3 \dfrac{\ell^2}{6} + \dfrac{w\ell^3}{24EI}$

$\dfrac{d^2 y}{dx^2} = 0$ at $x = \ell$ gives $0 = y_3 + \dfrac{w\ell}{2EI}$

$\therefore y_3 = \dfrac{-w\ell}{2EI}$ and $y_1 = \dfrac{w\ell^3}{24EI}$

$y = \dfrac{wx}{24EI}(\ell^3 - 2\ell x^2 + x^3)$

Of course, this d.e. can be solved by direct integration, but it has been included here to illustrate the L.T. treatment of an equation which (i) is of order higher than the second, and (ii) has boundary conditions which are not all initial conditions. A more complicated d.e. would incur a lot of tedious algebra, which is not rewarding from the student's point of view.

FRAME 40 continued

7. (i) $s^2\bar{x} + 2ks\bar{x} + \omega^2\bar{x} = \dfrac{s+k}{(s+k)^2 + p^2}$

$\bar{x} = \dfrac{s+k}{\{(s+k)^2 + n^2\}\{(s+k)^2 + p^2\}}$

The R.H.S. is of a similar form to $\dfrac{s}{(s^2+n^2)(s^2+p^2)}$ with $(s+k)$ replacing s, and $\mathcal{L}^{-1}\left\{\dfrac{s}{(s^2+n^2)(s^2+p^2)}\right\}$ can be written down by referring to your solution to 5(i).

Hence $x = \dfrac{e^{-kt}}{n^2 - p^2}(\cos pt - \cos nt)$

(ii) $\bar{x} = \dfrac{s+k}{\{(s+k)^2 + n^2\}^2}$

$x = \dfrac{1}{2n} e^{-kt} t \sin nt$

– SOLUTION BY LAPLACE TRANSFORM METHODS

APPENDIX

$$F(s) = \mathcal{L}\{f(t)\} = \int_0^\infty e^{-st} f(t)\, dt$$

TABLE OF LAPLACE TRANSFORM PAIRS

$f(t)$	$F(s)$
0	0
1	$1/s$
k	k/s
t	$1/s^2$
t^n	$n!/s^{n+1}$
$e^{\alpha t}$	$1/(s - \alpha)$
$\sin \omega t$	$\omega/(s^2 + \omega^2)$
$\cos \omega t$	$s/(s^2 + \omega^2)$
$\sin(\omega t + \phi)$	$(s \sin \phi + \omega \cos \phi)/(s^2 + \omega^2)$
$\cos(\omega t + \phi)$	$(s \cos \phi - \omega \sin \phi)/(s^2 + \omega^2)$
$\sinh \beta t$	$\beta/(s^2 - \beta^2)$
$\cosh \beta t$	$s/(s^2 - \beta^2)$
$\dfrac{1}{a - b}(e^{at} - e^{bt})$	$1/(s - a)(s - b)$
$\dfrac{1}{a - b}(ae^{at} - be^{bt})$	$s/(s - a)(s - b)$
$1 - e^{\alpha t}$	$-\alpha/s(s - \alpha)$
$1 - \cos \omega t$	$\omega^2/s(s^2 + \omega^2)$
$e^{\alpha t} f(t)$	$F(s - \alpha)$
$e^{\alpha t} t^n$	$n!/(s - \alpha)^{n+1}$
$e^{\alpha t} \sin \omega t$	$\omega/\{(s - \alpha)^2 + \omega^2\}$

APPENDIX continued

TABLE OF LAPLACE TRANSFORM PAIRS - continued

$f(t)$	$F(s)$
$e^{\alpha t} \cos \omega t$	$(s - \alpha)/\{(s - \alpha)^2 + \omega^2\}$
$t \sin \omega t$	$2\omega s/(s^2 + \omega^2)^2$
$t \cos \omega t$	$(s^2 - \omega^2)/(s^2 + \omega^2)^2$
$\sin \omega t - \omega t \cos \omega t$	$2\omega^3/(s^2 + \omega^2)^2$
$\dfrac{dx}{dt}$	$s\bar{x} - x_o$
$\dfrac{d^2x}{dt^2}$	$s^2\bar{x} - sx_o - x_1$
$\dfrac{d^nx}{dt^n}$	$s^n\bar{x} - s^{n-1}x_o - s^{n-2}x_1 - \ldots - x_{n-1}$

SIMULTANEOUS DIFFERENTIAL EQUATIONS

– SOLUTION BY D-OPERATOR AND LAPLACE TRANSFORM METHODS

A PROGRAMMED TEXT

A. C. Bajpai
I. M. Calus

INSTRUCTIONS

This programme constitutes a self-instructional course on the solution of simultaneous differential equations by D-operator and Laplace Transform methods.

The programme is divided up into a number of FRAMES which are to be worked *in the order given*. You will be required to participate in many of these frames and in such cases the answers are provided in ANSWER FRAMES, designated by the letter A following the frame number. Steps in the working are given where this is considered helpful. The answer frame is separated from the main frame by a line of asterisks: ******************. Keep the answers covered until you have written your own response. If your answer is wrong, go back and try to see why. Do not proceed to the next frame until you have corrected any mistakes in your attempt and are satisfied that you understand the contents up to this point.

Before reading this programme it is necessary that you are familiar with the following

Prerequisites

For the D-operator section — the contents of Programme 3.

For the Laplace Transform section — the contents of Programme 4.

CONTENTS

Instructions

FRAMES

1 - 4	Introduction
5 - 15	Examples solved by D-operator method
16	Remarks on the number of arbitrary constants
17 - 18	Applications using D-operator method
19	Branch instruction
20 - 30	Examples solved by Laplace Transform method
31 - 32	Applications using Laplace Transform method
33	Advantages of Laplace Transform method
34	Miscellaneous Examples
35	Answers to Miscellaneous Examples
APPENDIX	Table of Laplace Transform pairs

SOLUTION BY D-OPERATOR AND LAPLACE TRANSFORM METHODS

FRAME 1

Introduction

In various problems in science and engineering - for example, in mechanical and electrical coupled oscillations - more than one differential equation is required to describe the system.

We will first consider the oscillation of two weights suspended one below the other by springs.

Let their masses be m_1 and m_2, and suppose them suspended from a support at A by springs of stiffness k_1 and k_2 respectively. The distances at time t of m_1 and m_2 below their respective equilibrium positions (E.P.) are x_1 and x_2.

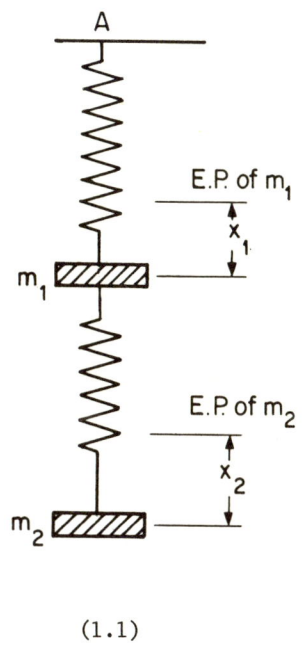

The equation of motion for m_1 is

$$m_1 \frac{d^2 x_1}{dt^2} = k_2(x_2 - x_1) - k_1 x_1$$

and that for m_2 is

$$m_2 \frac{d^2 x_2}{dt^2} = - k_2(x_2 - x_1).$$

The above equations can be written as:

$$\left. \begin{array}{l} m_1 \dfrac{d^2 x_1}{dt^2} + (k_1 + k_2)x_1 - k_2 x_2 = 0 \\[2ex] m_2 \dfrac{d^2 x_2}{dt^2} + k_2 x_2 - k_2 x_1 = 0 \end{array} \right\} \quad (1.1)$$

To obtain the positions of m_1 and m_2 at time t, it is necessary to solve these equations for x_1 and x_2.

Furthermore, if a periodic disturbing force is introduced (e.g. if A is moved up and down with simple harmonic motion) the equations then become:

$$\left. \begin{array}{l} m_1 \dfrac{d^2 x_1}{dt^2} + (k_1 + k_2)x_1 - k_2 x_2 = a \sin \omega t \\[2ex] m_2 \dfrac{d^2 x_2}{dt^2} + k_2 x_2 - k_2 x_1 = 0 \end{array} \right\} \quad (1.2)$$

FRAME 2

For our second illustration, we consider coupled electric circuits. Let one circuit (the primary) have a resistance R_1, an inductance L_1 and an e.m.f. e_1 and let the other circuit (the secondary) have a resistance R_2, an inductance L_2 and an e.m.f. e_2. The mutual inductance of the coils is M and the currents at time t in the two circuits are i_1 and i_2 respectively.

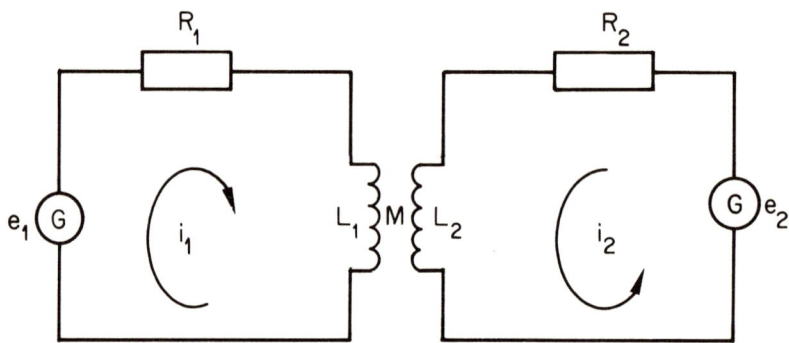

By Ohm's law, we get for the first circuit:

$$R_1 i_1 = e_1 - L_1 \frac{di_1}{dt} - M \frac{di_2}{dt}.$$

Similarly, for the second circuit we have:

$$R_2 i_2 = e_2 - L_2 \frac{di_2}{dt} - M \frac{di_1}{dt}.$$

Rearranging the above equations we get:

$$\left. \begin{array}{l} L_1 \dfrac{di_1}{dt} + R_1 i_1 + M \dfrac{di_2}{dt} = e_1 \\[2ex] M \dfrac{di_1}{dt} + L_2 \dfrac{di_2}{dt} + R_2 i_2 = e_2 \end{array} \right\} \quad (2.1)$$

The e.m.f.'s e_1 and e_2 could be constant or sinusoidal.

On solution, equations (2.1) give the currents i_1, i_2 at time t.

SOLUTION BY D-OPERATOR AND LAPLACE TRANSFORM METHODS

FRAME 3

In the earlier programmes in this series we have shown how to obtain the solution $y = f(x)$ to such differential equations as $\frac{dy}{dx} = F(x,y)$ and $a\frac{d^2y}{dx^2} + b\frac{dy}{dx} + cy = Q$. You will note that in these cases there was <u>one</u> variable y to be found in terms of the independent variable x.

In equations (1.2) in FRAME 1 there are <u>two</u> variables x_1 and x_2 which are to be found in terms of the single independent variable t.

Similarly in equations (2.1) of FRAME 2, both i_1 and i_2 have to be found as functions of t. In both examples, there are <u>two</u> differential equations which are necessary for finding the <u>two</u> unknown variables.

Such pairs of differential equations are called SIMULTANEOUS DIFFERENTIAL EQUATIONS. (Obviously the number of equations is not limited to two – for example a problem might give rise to <u>three</u> simultaneous d.e.'s from which <u>three</u> currents i_1, i_2, i_3 are to be found as functions of t. Cases of this kind will not be considered in this programme as they do not involve any different theory, only more complicated working.)

Note that in equations (1.2) and (2.1) all the coefficients are constants. In this programme we shall restrict ourselves to the solution of simultaneous differential equations with constant coefficients.

FRAME 4

In this programme we shall discuss the solution of simultaneous differential equations by the D-operator and Laplace Transform techniques. In each case we shall assume that you are familiar with the earlier programmes in which these techniques were introduced. The programme is so arranged that, if you wish, you can study either or both of these two methods.

D-operator methods will now be dealt with in FRAMES 5 – 19.

For Laplace Transform methods, see FRAMES 20 – 33.

Miscellaneous examples for both methods can be found at the end of the programme.

SIMULTANEOUS DIFFERENTIAL EQUATIONS –

FRAME 5

Solution by D-operator method

The principles underlying this method were explained in the earlier programme in this series – "Differential Equations with Constant Coefficients – Solution by D-operator Methods". We shall show you how to solve simultaneous d.e.'s by this method by considering a few examples.

Example 1

Solve for x and y, each as a function of t, the following d.e.'s:

$$\left. \begin{array}{l} \dfrac{dx}{dt} - 2x - 3y = 0 \\[6pt] \dfrac{dy}{dt} + x + 2y = 0 \end{array} \right\} \qquad (5.1)$$

Using $D \equiv \dfrac{d}{dt}$, equations (5.1) may be rewritten as:

$$Dx - 2x - 3y = 0$$
$$Dy + x + 2y = 0$$

Rearrangement then gives:

$$(D - 2)x - 3y = 0 \qquad (5.2)$$
$$x + (D + 2)y = 0 \qquad (5.3)$$

As you know, D obeys the fundamental laws of algebra. Hence these equations can be treated in the same way as a set of simultaneous algebraic equations. We can solve them by first eliminating either x or y. In this example, it is simpler to eliminate x. Operating on (5.3) by (D − 2), i.e. multiplying throughout by (D − 2):

$$(D - 2)x + (D^2 - 4)y = 0 \qquad (5.4)$$

Subtracting (5.2) from (5.4):

$$(D^2 - 1)y = 0$$

This is now a second order d.e. in y <u>only</u>. Solve for y.

****

SOLUTION BY D-OPERATOR AND LAPLACE TRANSFORM METHODS

5A

The A.E. is $m^2 - 1 = 0$
$$m = \pm 1.$$

Hence $y = Ae^t + Be^{-t}$ where A and B are arbitrary constants.

FRAME 6

x can then be found either by eliminating y in a similar manner or by substituting the solution for y in one of the equations. In this example, it will be simpler to substitute for y in (5.3).

This gives
$$\begin{aligned} x &= -(D + 2)y \\ &= -(D + 2)(Ae^t + Be^{-t}) \\ &= -(Ae^t - Be^{-t} + 2Ae^t + 2Be^{-t}) \\ &= -(3Ae^t + Be^{-t}). \end{aligned}$$

The general solution of equations (5.1) is therefore
$$x = -(3Ae^t + Be^{-t})$$
$$y = Ae^t + Be^{-t}.$$

You will note that in the solution of these <u>two</u> simultaneous d.e.'s of the <u>first</u> order, there are only <u>two</u> arbitrary constants.

FRAME 7

Example 2

Solve the equations

$$\left. \begin{aligned} \frac{dx}{dt} - x - 5y &= 1 \\ \frac{dy}{dt} + 2x + y &= e^t \end{aligned} \right\} \quad (7.1)$$

subject to the conditions that $x = y = 0$ at $t = 0$.

FRAME 7 continued

Rewriting the equations (7.1) gives:

$$(D - 1)x - 5y = 1 \qquad (7.2)$$
$$2x + (D + 1)y = e^t \qquad (7.3)$$

Either x or y can be eliminated. We shall eliminate x.

Multiplying (7.2) by 2 and operating on (7.3) by (D - 1), and then subtracting:

$$(D^2 + 9)y = (D - 1)e^t - 2$$
$$= e^t - e^t - 2$$
$$\therefore (D^2 + 9)y = -2$$

C.F. is $A \cos 3t + B \sin 3t$.

P.I. is $\dfrac{1}{D^2 + 9}(-2) = \dfrac{1}{9(1 + \frac{D^2}{9})}(-2)$

$$= \frac{1}{9}\left(1 + \frac{D^2}{9}\right)^{-1}(-2)$$

$$= -\frac{2}{9}.$$

(In case of difficulty, refer to FRAME 16 in the earlier programme on the D-operator.)

Hence $y = A \cos 3t + B \sin 3t - \dfrac{2}{9}.$ \qquad (7.4)

Now substitute this result in (7.3) to obtain x, as in FRAME 6.

Then apply the given initial conditions to find the particular solutions for x and y as required.

****

7A

$$2x = e^t - (D + 1)y$$
$$= e^t + 3A \sin 3t - 3B \cos 3t - A \cos 3t - B \sin 3t + \frac{2}{9}$$
$$x = \frac{1}{2}e^t + \frac{1}{9} + \frac{1}{2}(3A - B) \sin 3t - \frac{1}{2}(A + 3B) \cos 3t$$

When $t = 0$, $x = 0$ and $y = 0$. Substituting these we get

SOLUTION BY D-OPERATOR AND LAPLACE TRANSFORM METHODS

7A continued

$$0 = \frac{1}{2} + \frac{1}{9} - \frac{A}{2} - \frac{3B}{2}$$

$$\text{and } 0 = A - \frac{2}{9}$$

giving $A = \frac{2}{9} \quad B = \frac{1}{3}$

Hence $x = \frac{1}{2}e^t + \frac{1}{9} + \frac{1}{6}\sin 3t - \frac{11}{18}\cos 3t$

$y = \frac{2}{9}\cos 3t + \frac{1}{3}\sin 3t - \frac{2}{9}.$

FRAME 8

Example 3

Solve the following d.e.'s for x and y:

$$(3D - 2)x + Dy = 3\sin t + 5\cos t \qquad (8.1)$$
$$2Dx + (D + 1)y = \sin t + \cos t. \qquad (8.2)$$

We first eliminate y by operating on (8.1) by (D + 1) and on (8.2) by D, and then subtracting.

$$\{(D + 1)(3D - 2) - 2D^2\}x = (D + 1)(3\sin t + 5\cos t) - D(\sin t + \cos t)$$
$$(D^2 + D - 2)x = 7\cos t - \sin t$$

Now solve for x.

****

8A

C.F. is $Ae^t + Be^{-2t}$ where A and B are arbitrary constants.

P.I. is $\dfrac{1}{D^2 + D - 2}(7\cos t - \sin t)$

$= \dfrac{7}{D - 3}\cos t - \dfrac{1}{D - 3}\sin t$

$= \dfrac{7(D + 3)}{-1 - 9}\cos t - \dfrac{D + 3}{-1 - 9}\sin t$

$= -\dfrac{7}{10}(-\sin t + 3\cos t) + \dfrac{1}{10}(\cos t + 3\sin t)$

$= -2\cos t + \sin t$

$x = -2\cos t + \sin t + Ae^t + Be^{-2t}$

FRAME 9

In Examples 1 and 2 we found the second solution by substitution. In those examples $\frac{dx}{dt}$ and $\frac{dy}{dt}$ did not both occur in the same equation. Hence finding the second solution by substitution was simpler. In the present example substituting the solution for x into either (8.1) or (8.2) would not give y immediately but would lead to a differential equation in y, which would then have to be solved for y. To avoid this, we could obtain another d.e. which does <u>not</u> contain Dy by eliminating Dy between (8.1) and (8.2). We can then find y from this new d.e. by the substitution method as applied in Examples 1 and 2.

Hence, subtracting (8.2) from (8.1) to eliminate Dy, we obtain the new d.e.

$$(D - 2)x - y = 2\sin t + 4\cos t \qquad (9.1)$$

or
$$y = (D - 2)x - 2\sin t - 4\cos t$$
$$= \cos t - 2\sin t - Ae^t - 4Be^{-2t} \quad \text{on substituting for x.}$$

There may be cases when you are required to solve simultaneous d.e.'s from which the new d.e. of the type (9.1) cannot be obtained. In such cases you could use the alternative method of elimination. This technique will now be demonstrated with the present example.

First eliminate x from equations (8.1) and (8.2) to obtain

$$(D^2 + D - 2)y = 5(\sin t - \cos t) \qquad (9.2)$$

****

9A

Operate on (8.1) by 2D and on (8.2) by (3D − 2) and subtract to obtain

$$\{2D^2 - (3D - 2)(D + 1)\}y = 6\cos t - 10\sin t - 3\cos t + 3\sin t$$
$$+ 2\sin t + 2\cos t$$

giving $\quad (D^2 + D - 2)y = 5(\sin t - \cos t)$

SOLUTION BY D-OPERATOR AND LAPLACE TRANSFORM METHODS

FRAME 10

We now solve (9.2) for y.

The A.E. is $\quad m^2 + m - 2 = 0$

$$m = 1, -2$$

∴ C.F. is $Ce^t + Ee^{-2t}$ where C and E are arbitrary constants which are not independent of A and B. We shall show later how C and E are related to A and B. We only require <u>two</u> arbitrary constants (see FRAME 6).
Find the P.I.

****

10A

$$P.I. \text{ is } \frac{1}{D^2 + D - 2} 5(\sin t - \cos t)$$

$$= 5\{\frac{D + 3}{-10} \sin t - \frac{D + 3}{-10} \cos t\}$$

$$= \cos t - 2 \sin t.$$

FRAME 11

Now we can write the complete solution for y.

$$y = \cos t - 2 \sin t + Ce^t + Ee^{-2t}.$$

From FRAME 8A we know that $x = -2 \cos t + \sin t + Ae^t + Be^{-2t}$.
To find C and E in terms of A and B, we substitute for x and y in either (8.1) or (8.2).
Substitution in (8.2) gives

$$4 \sin t + 2 \cos t + 2Ae^t - 4Be^{-2t}$$
$$- \sin t - 2 \cos t + Ce^t - 2Ee^{-2t}$$
$$- 2 \sin t + \cos t + Ce^t + Ee^{-2t} = \sin t + \cos t$$

or $\quad (2A + 2C)e^t - (4B + E)e^{-2t} = 0.$

Equating coefficients we get $\quad C = -A$
$$E = -4B.$$

Hence the solution is

$$x = -2 \cos t + \sin t + Ae^t + Be^{-2t}$$
$$y = \cos t - 2 \sin t - Ae^t - 4Be^{-2t}.$$

If two boundary conditions were given, A and B could then be found.

FRAME 12

Example 4

Solve the following d.e.'s for x and y:

$$\frac{d^2x}{dt^2} - 2x + 6y = \frac{3}{2} \sin 2t \qquad (12.1)$$

$$\frac{d^2y}{dt^2} + x - y = 0. \qquad (12.2)$$

Rewriting these equations we get

$$(D^2 - 2)x + 6y = \frac{3}{2} \sin 2t$$

$$x + (D^2 - 1)y = 0.$$

Eliminate x to obtain a d.e. for y.

****

12A

$$\{(D^2 - 2)(D^2 - 1) - 6\}y = -\frac{3}{2} \sin 2t$$

$$(D^4 - 3D^2 - 4)y = -\frac{3}{2} \sin 2t$$

FRAME 13

We can solve for y from the d.e.

$$(D^4 - 3D^2 - 4)y = -\frac{3}{2} \sin 2t$$

$$(D^2 - 4)(D^2 + 1)y = -\frac{3}{2} \sin 2t.$$

The C.F. is $Ae^{2t} + Be^{-2t} + C \cos t + E \sin t$.

The P.I. is $\dfrac{1}{(D^2 - 4)(D^2 + 1)} \left(-\dfrac{3}{2} \sin 2t\right)$

$$= \frac{-3}{2(-8)(-3)} \sin 2t$$

$$= -\frac{1}{16} \sin 2t$$

$\therefore \quad y = -\dfrac{1}{16} \sin 2t + Ae^{2t} + Be^{-2t} + C \cos t + E \sin t.$

x could now be found either by substitution or by elimination, as illustrated

SOLUTION BY D-OPERATOR AND LAPLACE TRANSFORM METHODS

FRAME 13 continued

in the previous examples. In this example we recommend substitution in (12.2) which leads to the solution for x directly. Complications arise in the elimination method due to the necessity to introduce four more arbitrary constants which then have to be related to A,B,C and E.

Now solve for x.

****

13A

From (12.2) $\quad x = (1 - D^2)y$

$$= -\frac{1}{16} \sin 2t + Ae^{2t} + Be^{-2t} + C \cos t + E \sin t$$
$$\quad - (\frac{1}{4} \sin 2t + 4Ae^{2t} + 4Be^{-2t} - C \cos t - E \sin t)$$
$$= -\frac{5}{16} \sin 2t - 3Ae^{2t} - 3Be^{-2t} + 2C \cos t + 2E \sin t.$$

FRAME 14

Note that in Example 4, the solution of the <u>two</u> simultaneous d.e.'s of the <u>second order</u> contained only <u>four</u> arbitrary constants.

In the following example, we shall show that this is not always true.

Example 5

Solve for x and y:

$$(D^2 + 2)x + D^2 y = 3 \cos 2t \qquad (14.1)$$
$$D^2 x + (D^2 - 1)y = 0. \qquad (14.2)$$

First solve for y by eliminating x.

****

14A

$$\{D^4 - (D^2 + 2)(D^2 - 1)\}y = -12 \cos 2t$$
$$\textit{yielding} \qquad (D^2 - 2)y = 12 \cos 2t$$

                                                                    14A continued

The C.F. is $Ae^{\sqrt{2}t} + Be^{-\sqrt{2}t}$.

The P.I. is $\dfrac{1}{D^2 - 2} 12 \cos 2t = -2 \cos 2t$.

$$y = -2 \cos 2t + Ae^{\sqrt{2}t} + Be^{-\sqrt{2}t}.$$

Note that there are only two *arbitrary constants in the solution.*

                                                                    FRAME 15

As before, x can be found by substitution or elimination. Direct substitution in (14.2) will give D^2x, from which x can be found by integrating twice, involving two constants of integration which will have to be determined. However, as in FRAME 9, we could obtain a new equation by eliminating D^2x between (14.1) and (14.2) and then find x by substituting for y. Perhaps you would like to try at least one of these methods.

****
                                                                    15A

In this example, the simplest method is to eliminate D^2x.
This gives $2x + y = 3 \cos 2t$.

$$x = \tfrac{1}{2}(3 \cos 2t - y)$$
$$= -\tfrac{A}{2} e^{\sqrt{2}t} - \tfrac{B}{2} e^{-\sqrt{2}t} + \tfrac{5}{2} \cos 2t$$

Direct substitution in (14.2) gives

$$D^2x = (1 - D^2)y$$
$$= -Ae^{\sqrt{2}t} - Be^{-\sqrt{2}t} - 10 \cos 2t.$$

On integration, $Dx = -\dfrac{A}{\sqrt{2}} e^{\sqrt{2}t} + \dfrac{B}{\sqrt{2}} e^{-\sqrt{2}t} - 5 \sin 2t + C.$

Integrating again, $x = -\dfrac{A}{2} e^{\sqrt{2}t} - \dfrac{B}{2} e^{-\sqrt{2}t} + \dfrac{5}{2} \cos 2t + Ct + E.$

SOLUTION BY D-OPERATOR AND LAPLACE TRANSFORM METHODS   5:13

<u>15A</u> continued

This must obviously satisfy (14.2) for all C and E. We now substitute for x and y in (14.1). Since the only terms on the L.H.S. which involve C and E are $Ct + E$ and as the R.H.S. contains no t or constant terms, both C and E must be zero.

$$x = -\frac{A}{2} e^{\sqrt{2}t} - \frac{B}{2} e^{-\sqrt{2}t} + \frac{5}{2} \cos 2t$$

Alternatively, the elimination method will yield

$$\{(D^2 - 1)(D^2 + 2) - D^4\}x = 3(D^2 - 1)\cos 2t$$

or

$$(D^2 - 2)x = -15 \cos 2t$$

from which $x = Ce^{\sqrt{2}t} + Ee^{-\sqrt{2}t} + \frac{5}{2} \cos 2t$ where C and E are arbitrary constants.

If we now substitute for x and y in either of the two d.e.'s, preferably (14.2), we get

$$2Ce^{\sqrt{2}t} + 2Ee^{-\sqrt{2}t} - 10 \cos 2t$$
$$+ 2Ae^{\sqrt{2}t} + 2Be^{-\sqrt{2}t} + 8 \cos 2t$$
$$- Ae^{\sqrt{2}t} - Be^{-\sqrt{2}t} + 2 \cos 2t = 0.$$

Equating the coefficients on both sides gives

$$2C + A = 0 \quad \text{and} \quad 2E + B = 0.$$

$$C = -\frac{A}{2} \qquad E = -\frac{B}{2}$$

$$\therefore \quad x = -\frac{A}{2} e^{\sqrt{2}t} - \frac{B}{2} e^{-\sqrt{2}t} + \frac{5}{2} \cos 2t.$$

FRAME 16

<u>Remarks on the number of arbitrary constants.</u>

The maximum number of arbitrary constants is equal to the sum of the orders of the differential equations (as in Examples 1 - 4). There may be less (as in Example 5). This depends on the highest power of D in the eliminant, e.g. in Example 5 the eliminant is

FRAME 16 continued

$$\begin{vmatrix} D^2 + 2 & D^2 \\ D^2 & D^2 - 1 \end{vmatrix} = D^2 - 2$$

The highest power of D here is 2, so there should be only <u>two</u> constants. How many constants do you expect in the solution of the following simultaneous d.e.'s?

$$(D^2 - 1)x + (D^2 - D + 1)y = \sin 2t$$
$$D^2 x + (D^2 + 3)y = 0.$$

****

16A

Three arbitrary constants, since in

$$\begin{vmatrix} D^2 - 1 & D^2 - D + 1 \\ D^2 & D^2 + 3 \end{vmatrix}$$

the highest power of D is 3.

FRAME 17

We shall now give two examples of the solution, by D-operator methods, of simultaneous d.e.'s arising in physical problems.

<u>Application 1</u>

We first consider the oscillation of two weights suspended one below the other by springs, as described in FRAME 1.

In such a system, if $m_1 = 18$, $m_2 = 3$, $k_1 = 108$ and $k_2 = 18$ the simultaneous d.e.'s (1.1) become

$$18 \frac{d^2 x_1}{dt^2} + 126 x_1 - 18 x_2 = 0$$
$$3 \frac{d^2 x_2}{dt^2} + 18 x_2 - 18 x_1 = 0$$

giving

$$\frac{d^2 x_1}{dt^2} + 7 x_1 - x_2 = 0$$
$$\frac{d^2 x_2}{dt^2} + 6 x_2 - 6 x_1 = 0.$$

Now use the D-operator method to find x_1 and x_2 in terms of t.

****

SOLUTION BY D-OPERATOR AND LAPLACE TRANSFORM METHODS

17A

In D-operator form the equations become

$$(D^2 + 7)x_1 - x_2 = 0 \qquad (17A.1)$$
$$-6x_1 + (D^2 + 6)x_2 = 0. \qquad (17A.2)$$

Eliminating x_2, we get

$$\{(D^2 + 6)(D^2 + 7) - 6\}x_1 = 0$$
$$(D^2 + 4)(D^2 + 9)x_1 = 0.$$
$$x_1 = A \cos 2t + B \sin 2t + C \cos 3t + E \sin 3t.$$

Substitution of x_1 in (17A.1) gives

$$x_2 = (D^2 + 7)(A \cos 2t + B \sin 2t + C \cos 3t + E \sin 3t)$$
$$= 3A \cos 2t + 3B \sin 2t - 2C \cos 3t - 2E \sin 3t.$$

FRAME 18

Application 2

For the second application we take the case of coupled electric circuits, which was discussed in FRAME 2.

In such a system, if $R_1 = R_2 = R$, $L_1 = L_2 = L$, $e_1 = E$ (a constant) and $e_2 = 0$, the simultaneous d.e.'s (2.1) become

$$L \frac{di_1}{dt} + Ri_1 + M \frac{di_2}{dt} = E$$
$$M \frac{di_1}{dt} + L \frac{di_2}{dt} + Ri_2 = 0.$$

Now use the D-operator method to find the currents i_1 and i_2 after time t, assuming that initially they are zero.

****

18A

In the D-operator form the equations become

$$(LD + R)i_1 + MDi_2 = E \qquad (18A.1)$$
$$MDi_1 + (LD + R)i_2 = 0. \qquad (18A.2)$$

Elimination of i_1 gives

18A continued

$$\{M^2D^2 - (LD + R)^2\}i_2 = 0$$
$$\{(M + L)D + R\}\{(M - L)D - R\}i_2 = 0$$

$$\therefore i_2 = Ae^{-\frac{Rt}{M+L}} + Be^{\frac{Rt}{M-L}}$$

$$= Ae^{-\frac{Rt}{L+M}} + Be^{-\frac{Rt}{L-M}}. \qquad (18A.3)$$

Substitution of i_2 in (18A.2) yields

$$MDi_1 = -(LD + R)\left(Ae^{-\frac{Rt}{L+M}} + Be^{-\frac{Rt}{L-M}}\right)$$

$$= \left(\frac{LR}{L+M} - R\right)Ae^{-\frac{Rt}{L+M}} + \left(\frac{LR}{L-M} - R\right)Be^{-\frac{Rt}{L-M}}$$

$$Di_1 = -\frac{AR}{L+M}e^{-\frac{Rt}{L+M}} + \frac{BR}{L-M}e^{-\frac{Rt}{L-M}}.$$

On integration, $i_1 = Ae^{-\frac{Rt}{L+M}} - Be^{-\frac{Rt}{L-M}} + K$ (a constant).

To find K, substitute in (18A.1).

Comparing coefficients, (it is only necessary to consider the constant terms),

$$RK = E \quad giving \quad K = E/R$$

$$\therefore i_1 = Ae^{-\frac{Rt}{L+M}} - Be^{-\frac{Rt}{L-M}} + \frac{E}{R}. \qquad (18A.4)$$

We know that $i_1 = i_2 = 0$ at $t = 0$.

$$\therefore 0 = A + B \qquad from\ (18A.3)$$
$$0 = A - B + \frac{E}{R} \qquad from\ (18A.4).$$

Solving, $A = -B = -E/2R$.

$$\therefore i_1 = \frac{E}{2R}\left\{2 - e^{-\frac{Rt}{L+M}} - e^{-\frac{Rt}{L-M}}\right\}$$

$$i_2 = \frac{E}{2R}\left\{e^{-\frac{Rt}{L-M}} - e^{-\frac{Rt}{L+M}}\right\}.$$

SOLUTION BY D-OPERATOR AND LAPLACE TRANSFORM METHODS

FRAME 19

As indicated in FRAME 4, you can now proceed directly to the Miscellaneous Examples in FRAME 34 if you do not wish to study the Laplace Transform method.

FRAME 20

Solution by Laplace Transform Method

The principles underlying this method were explained in the earlier programme in this series – "Differential Equations with Constant Coefficients – Solution by Laplace Transform Methods". We will show you how to solve simultaneous d.e.'s by this method by considering a few examples.

Example 1

Solve for x and y, each as a function of t, the following d.e.'s:

$$\left. \begin{array}{l} \dfrac{dx}{dt} - 2x - 3y = 0 \\[4pt] \dfrac{dy}{dt} + x + 2y = 0 \end{array} \right\} \qquad (20.1)$$

subject to the conditions $x = 2$, $y = 0$ at $t = 0$.

(NOTE: This pair of d.e.'s was solved in FRAME 5 by D-operator. You will recall that Laplace Transform methods are especially suitable for problems with initial conditions. We have therefore supplied initial conditions.)

Transforming equations (20.1) and writing $\bar{x}, \bar{y}$ for $\mathcal{L}\{x\}$ and $\mathcal{L}\{y\}$ respectively, we get

$$(s\bar{x} - 2) - 2\bar{x} - 3\bar{y} = 0$$
$$(s\bar{y} - 0) + \bar{x} + 2\bar{y} = 0.$$

Rearranging we get

$$\left. \begin{array}{l} (s - 2)\bar{x} - 3\bar{y} = 2 \\ \bar{x} + (s + 2)\bar{y} = 0. \end{array} \right\} \qquad (20.2)$$

These are simultaneous algebraic equations in $\bar{x}$ and $\bar{y}$. Eliminating $\bar{x}$ gives

$$(s^2 - 4 + 3)\bar{y} = -2$$
$$\bar{y} = \dfrac{-2}{s^2 - 1}$$

SIMULTANEOUS DIFFERENTIAL EQUATIONS -

FRAME 20 continued

Inversion gives $y = -2 \sinh t$.   (See Table of Laplace Transform Pairs in APPENDIX.)

This solution could also be expressed as

$$y = e^{-t} - e^{t}.$$

FRAME 21

x can then be found either by eliminating $\bar{y}$ between equations (20.2) or by substituting the solution for y in one of the equations (20.1). In this example, it will be simpler to substitute for y in the second equation of (20.1).

This gives

$$x = -\frac{dy}{dt} - 2y$$
$$= e^{-t} + e^{t} - 2e^{-t} + 2e^{t}$$
$$= 3e^{t} - e^{-t}.$$

The required solution is, therefore,

$$x = 3e^{t} - e^{-t}$$
$$y = e^{-t} - e^{t}.$$

You will recall that Laplace Transform methods incorporate the initial conditions in the solution at the outset, without having to find a general solution containing arbitrary constants.

FRAME 22

In the example which we have just solved, we could have used the Laplace Transform method even if initial conditions were not given. In such a case we would denote the values of x and y at $t = 0$ by x_0 and y_0 respectively, and proceed as before.

For practice, solve for x and y equations (20.1) without any specific initial conditions being given.

****

SOLUTION BY D-OPERATOR AND LAPLACE TRANSFORM METHODS

22A

Let $x = x_o$ and $y = y_o$ at $t = 0$.

The transformed equations will be

$$(s\bar{x} - x_o) - 2\bar{x} - 3\bar{y} = 0$$

$$(s\bar{y} - y_o) + \bar{x} + 2\bar{y} = 0.$$

Rearranging and eliminating $\bar{x}$ will give

$$\bar{y} = \frac{(s-2)y_o - x_o}{s^2 - 1}$$

$$= \frac{sy_o}{s^2 - 1} - \frac{2y_o + x_o}{s^2 - 1}.$$

Inversion gives $y = y_o \cosh t - (2y_o + x_o) \sinh t$.

x can be obtained either by elimination of $\bar{y}$ or by substitution, as before, giving

$$x = x_o \cosh t + (3y_o + 2x_o) \sinh t.$$

Needless to say, the above solutions could easily be expressed in exponential form.

FRAME 23

Example 2

Solve the equations

$$\frac{dx}{dt} - x - 5y = 1$$

$$\frac{dy}{dt} + 2x + y = e^t$$

subject to the conditions that $x = y = 0$ at $t = 0$.
(NOTE: This example was solved by D-operator in FRAME 7.)

Transform the equations and eliminate $\bar{x}$ to find $\bar{y}$, and hence y.

****

  23A

The transformed equations, after rearranging, are

$$(s - 1)\bar{x} - 5\bar{y} = \frac{1}{s}$$

$$2\bar{x} + (s + 1)\bar{y} = \frac{1}{s - 1}.$$

Eliminating $\bar{x}$ gives $(s^2 + 9)\bar{y} = 1 - \frac{2}{s}$

$$\bar{y} = \frac{1}{s^2 + 9} - \frac{2}{s(s^2 + 9)}.$$

Inversion gives $y = \frac{1}{3} \sin 3t - \frac{2}{9}(1 - \cos 3t)$.

(NOTE: $\mathcal{L}^{-1}\{\frac{\omega^2}{s(s^2 + \omega^2)}\} = 1 - \cos \omega t$ *from the APPENDIX)*

                                                              FRAME 24

Having found y, now obtain x.

****

                                                                    24A

By far the simplest way of obtaining x is to substitute in the second d.e.

$$2x = e^t - y - \frac{dy}{dt}.$$

Thus $x = \frac{1}{2}(e^t + \frac{1}{3} \sin 3t - \frac{11}{9} \cos 3t + \frac{2}{9})$.

                                                              FRAME 25

Example 3

Solve the following d.e.'s for x and y:

$$(3D - 2)x + Dy = 3 \sin t + 5 \cos t \quad (25.1)$$
$$2Dx + (D + 1)y = \sin t + \cos t \quad (25.2)$$

where $D \equiv \frac{d}{dt}$, and $x = 0$, $y = -1$ when $t = 0$.

(NOTE: This pair of d.e.'s, without the initial conditions, was solved by D-operator in FRAMES 8 - 11.)

SOLUTION BY D-OPERATOR AND LAPLACE TRANSFORM METHODS

FRAME 25 continued

Of course, $\mathcal{L}\{Dx\} = \mathcal{L}\{\frac{dx}{dt}\}$ and $\mathcal{L}\{Dy\} = \mathcal{L}\{\frac{dy}{dt}\}$. Use these to transform the above d.e.'s, eliminate $\bar{y}$ and then find x.

****

25A

The transformed equations, on rearranging, are

$$(3s - 2)\bar{x} + s\bar{y} = \frac{3 + 5s}{s^2 + 1} - 1$$

$$2s\bar{x} + (s + 1)\bar{y} = \frac{1 + s}{s^2 + 1} - 1.$$

Eliminating $\bar{y}$ we get

$$\{(3s - 2)(s + 1) - 2s^2\}\bar{x} = \frac{(3 + 5s)(s + 1)}{s^2 + 1} - (s + 1) - \frac{s(1 + s)}{s^2 + 1} + s$$

yielding $(s + 2)(s - 1)\bar{x} = \frac{3s^2 + 7s + 2}{s^2 + 1}$

$$= \frac{(3s + 1)(s + 2)}{s^2 + 1}$$

$$\bar{x} = \frac{3s + 1}{(s^2 + 1)(s - 1)}$$

$$= \frac{2}{s - 1} + \frac{1 - 2s}{s^2 + 1} \quad \text{on resolving into partial fractions.}$$

$$\therefore \quad x = 2e^t + \sin t - 2 \cos t.$$

FRAME 26

y could be obtained either by eliminating $\bar{x}$ between the transformed equations or by substituting for x in (25.1). The latter will give Dy, from which y can be found by integration. The constant of integration can be determined by using the relevant initial conditions.
Sometimes (as in this example), another d.e. which does not contain Dy can be obtained by eliminating Dy between the original d.e.'s (25.1) and (25.2). We can then find y by substituting for x in this new d.e.
Now find x by any of these methods.

****

26A

In this example elimination of Dy between (25.1) and (25.2) gives a new d.e.

$$(D - 2)x - y = 2 \sin t + 4 \cos t$$

$$y = -2 \sin t - 4 \cos t + (D - 2)x$$

$$= \cos t - 2 \sin t - 2e^t.$$

Substituting for x in (25.1) would give

$$Dy = 3 \sin t + 5 \cos t - (3D - 2)x$$

$$= -\sin t - 2 \cos t - 2e^t$$

giving $\quad y = \cos t - 2 \sin t - 2e^t + C.$

Using $y = -1$, $t = 0$ *gives* $C = 0.$

$$y = \cos t - 2 \sin t - 2e^t.$$

Eliminating $\bar{x}$ *between the transformed equations will give*

$$\{(3s - 2)(s + 1) - 2s^2\}\bar{y} = \frac{(3s - 2)(1 + s)}{s^2 + 1} - (3s - 2) - \frac{(3 + 5s)2s}{s^2 + 1} + 2s$$

yielding $(s + 2)(s - 1)\bar{y} = \dfrac{-7s^2 - 5s - 2}{s^2 + 1} - s + 2.$

$$\bar{y} = \frac{-s(s^2 + 5s + 6)}{(s^2 + 1)(s + 2)(s - 1)}$$

$$= \frac{-s^2 - 3s}{(s^2 + 1)(s - 1)}$$

$$= \frac{-2}{s - 1} + \frac{s - 2}{s^2 + 1} \quad \text{on resolving into partial fractions}$$

$$y = \cos t - 2 \sin t - 2e^t.$$

SOLUTION BY D-OPERATOR AND LAPLACE TRANSFORM METHODS

FRAME 27

Example 4

Solve the following d.e.'s for x and y.

$$\frac{d^2x}{dt^2} - 2x + 6y = \frac{3}{2}\sin 2t$$

$$\frac{d^2y}{dt^2} + x - y = 0$$

subject to the conditions $x = y = 0$, $\frac{dx}{dt} = -1$, $\frac{dy}{dt} = 0$ at $t = 0$.

(NOTE: This pair of d.e.'s without the initial conditions, was solved by D-operator in FRAMES 12 - 13.)

Transform the above d.e.'s, eliminate $\bar{x}$ and then find y.

****

27A

The transformed equations, on rearranging, are

$$(s^2 - 2)\bar{x} + 6\bar{y} = \frac{3}{s^2 + 4} - 1$$

$$\bar{x} + (s^2 - 1)\bar{y} = 0.$$

Eliminating $\bar{x}$ we get

$$\{(s^2 - 2)(s^2 - 1) - 6\}\bar{y} = 1 - \frac{3}{s^2 + 4}.$$

$$(s^2 - 4)(s^2 + 1)\bar{y} = \frac{s^2 + 1}{s^2 + 4}$$

$$\bar{y} = \frac{1}{(s^2 + 4)(s^2 - 4)}$$

$$= \frac{1}{8}\left(\frac{1}{s^2 - 4} - \frac{1}{s^2 + 4}\right)$$

NOTE: $\frac{1}{(s^2 + 4)(s^2 - 4)}$ is a function of s^2 and you can therefore assume partial fractions of the form $\frac{A}{s^2 + 4} + \frac{B}{s^2 - 4}$. This is obvious if you put $u = s^2$ in the fraction, which then becomes $\frac{1}{(u + 4)(u - 4)}$ giving the partial fractions $\frac{A}{u + 4} + \frac{B}{u - 4}$.

Inversion gives $\quad y = \frac{1}{8}\{\frac{1}{4}(e^{2t} - e^{-2t}) - \frac{1}{2}\sin 2t\}$

$$= \frac{1}{32}(e^{2t} - e^{-2t} - 2\sin 2t).$$

SIMULTANEOUS DIFFERENTIAL EQUATIONS -

FRAME 28

Having found y, now obtain x.

****

28A

Substitution in the second d.e. of FRAME 27 gives
$$x = y - \frac{d^2y}{dt^2}$$
$$= \frac{1}{32}(-3e^{2t} + 3e^{-2t} - 10 \sin 2t).$$

x could also be found by eliminating $\bar{y}$ between the transformed equations given in 27A.

FRAME 29

Example 5

Solve the following d.e.'s for x and y
$$\frac{dx}{dt} + 2x + \frac{dy}{dt} + 3y = e^t \qquad (29.1)$$
$$\frac{dx}{dt} - x + 2\frac{dy}{dt} - y = 0 \qquad (29.2)$$

subject to the conditions $x = 0$, $y = 1$ at $t = 0$.

Transform the above d.e.'s, eliminate $\bar{x}$ and then find y.

****

29A

The transformed equations, on rearranging, are
$$(s + 2)\bar{x} + (s + 3)\bar{y} = \frac{s}{s - 1}$$
$$(s - 1)\bar{x} + (2s - 1)\bar{y} = 2$$

giving, on elimination,
$$\bar{y} = \frac{s + 4}{s^2 + s + 1} = \frac{s + \frac{1}{2}}{(s + \frac{1}{2})^2 + \left(\frac{\sqrt{3}}{2}\right)^2} + \frac{\frac{7}{2}}{(s + \frac{1}{2})^2 + \left(\frac{\sqrt{3}}{2}\right)^2}$$

On inversion, we get $y = e^{-\frac{1}{2}t}(\cos\frac{\sqrt{3}t}{2} + \frac{7}{\sqrt{3}}\sin\frac{\sqrt{3}t}{2}).$

SOLUTION BY D-OPERATOR AND LAPLACE TRANSFORM METHODS

FRAME 30

Now find x by eliminating $\bar{y}$, for further practice in this technique. (You could, of course, obtain x by eliminating $\frac{dx}{dt}$ between (29.1) and (29.2) and then substituting for y in the new d.e.)

****

30A

Elimination of $\bar{y}$ gives
$$\bar{x} = \frac{2s^2 - s}{(s - 1)(s^2 + s + 1)} - \frac{2(s + 3)}{s^2 + s + 1}.$$

The first fraction can be resolved into the partial fractions
$$\frac{1}{3}\frac{1}{s - 1} + \frac{1}{3}\frac{5s + 1}{s^2 + s + 1}.$$

$$\therefore \bar{x} = \frac{1}{3}\frac{1}{s - 1} - \frac{1}{3}\frac{s + 17}{s^2 + s + 1}$$

$$x = \frac{1}{3}e^t - \frac{1}{3}e^{-\frac{1}{2}t}\left[\cos\frac{\sqrt{3}t}{2} + 11\sqrt{3}\sin\frac{\sqrt{3}t}{2}\right].$$

FRAME 31

We shall now give two examples of the solutions of simultaneous d.e.'s arising in physical problems, by using Laplace Transform methods.
(If you followed the D-operator part of the programme, you will have dealt with these two applications by that technique in FRAMES 17 and 18.)

Application 1

We first consider the oscillation of two weights suspended one below the other by springs, as described in FRAME 1.

In such a system, if $m_1 = 18$, $m_2 = 3$, $k_1 = 108$ and $k_2 = 18$ the simultaneous d.e.'s (1.1) become

$$18\frac{d^2x_1}{dt^2} + 126x_1 - 18x_2 = 0$$

$$3\frac{d^2x_2}{dt^2} + 18x_2 - 18x_1 = 0$$

giving $\quad \dfrac{d^2x_1}{dt^2} + 7x_1 - x_2 = 0$

FRAME 31 continued

$$\frac{d^2x_2}{dt^2} + 6x_2 - 6x_1 = 0.$$

Now use the Laplace Transform method to find x_1 and x_2 in terms of t, taking as the initial conditions $x_1 = x_2 = 0$, $\dot{x}_1 = v$ (constant), $\dot{x}_2 = 0$ at $t = 0$.

****

31A

The transformed equations, on rearranging, are

$$(s^2 + 7)\bar{x}_1 - \bar{x}_2 = v$$

$$-6\bar{x}_1 + (s^2 + 6)\bar{x}_2 = 0.$$

Eliminating $\bar{x}_2$ we get

$$\{(s^2 + 6)(s^2 + 7) - 6\}\bar{x}_1 = v(s^2 + 6)$$

$$\bar{x}_1 = \frac{v(s^2 + 6)}{(s^2 + 9)(s^2 + 4)}$$

$$= \frac{2v}{5(s^2 + 4)} + \frac{3v}{5(s^2 + 9)}.$$

(Refer to NOTE in 27A for the simplest method of obtaining these P.F.'s.)

Inversion gives $x_1 = \frac{v}{5} \sin 2t + \frac{v}{5} \sin 3t.$

Substitution or elimination processes then give $x_2 = \frac{3v}{5} \sin 2t - \frac{2v}{5} \sin 3t.$

FRAME 32

Application 2

For the second application we take the case of coupled electric circuits, which was discussed in FRAME 2.

In such a system, if $R_1 = R_2 = R$, $L_1 = L_2 = L$, $e_1 = E$ (a constant) and $e_2 = 0$, the simultaneous d.e.'s (2.1) become

$$L \frac{di_1}{dt} + Ri_1 + M \frac{di_2}{dt} = E$$

$$M \frac{di_1}{dt} + L \frac{di_2}{dt} + Ri_2 = 0.$$

Now use the Laplace Transform method to find the currents i_1 and i_2 after time t, assuming that initially they are zero.

****

SOLUTION BY D-OPERATOR AND LAPLACE TRANSFORM METHODS

<u>32A</u>

The transformed equations, on rearranging, are

$$(Ls + R)\bar{i}_1 + Ms\bar{i}_2 = \frac{E}{s} \qquad (32A.1)$$

$$Ms\bar{i}_1 + (Ls + R)\bar{i}_2 = 0. \qquad (32A.2)$$

Eliminating $\bar{i}_2$ we get

$$\{(Ls + R)^2 - M^2s^2\}\bar{i}_1 = \frac{E(Ls + R)}{s}$$

$$\bar{i}_1 = E \frac{Ls + R}{s(Ls + Ms + R)(Ls - Ms + R)}$$

$$= \frac{EL}{(L + M)(L - M)} \cdot \frac{s + \frac{R}{L}}{s\left(s + \frac{R}{L + M}\right)\left(s + \frac{R}{L - M}\right)}.$$

The second fraction is of the form $\frac{s + \alpha}{s(s + \beta)(s + \gamma)}$ *which can be split into partial fractions of the form* $\frac{A}{s} + \frac{B}{s + \beta} + \frac{C}{s + \gamma}$, *giving*

$$A = \frac{\alpha}{\beta\gamma}, \qquad B = \frac{\alpha - \beta}{\beta(\beta - \gamma)} \quad \text{and} \quad C = \frac{\gamma - \alpha}{\gamma(\beta - \gamma)}.$$

$$\therefore \bar{i}_1 = E\left\{\frac{\frac{1}{R}}{s} - \frac{\frac{1}{2R}}{s + \frac{R}{L + M}} - \frac{\frac{1}{2R}}{s + \frac{R}{L - M}}\right\}, \quad \begin{array}{l}\textit{obtained after}\\ \textit{substituting and}\\ \textit{simplifying.}\end{array}$$

Inversion then gives

$$i_1 = \frac{E}{2R}\left(2 - e^{-\frac{Rt}{L+M}} - e^{-\frac{Rt}{L-M}}\right) \qquad (32A.3)$$

From (32A.2), $\bar{i}_2 = -\frac{Ms}{L(s + \frac{R}{L})}\bar{i}_1$

$$= -\frac{EM}{(L + M)(L - M)} \cdot \frac{1}{\left(s + \frac{R}{L + M}\right)\left(s + \frac{R}{L - M}\right)}$$

32A continued

Using the Laplace Transform pair in the APPENDIX:

$$\mathcal{L}^{-1}\left\{\frac{1}{(s-a)(s-b)}\right\} = \frac{1}{a-b}(e^{at} - e^{bt}),$$

we get
$$i_2 = \frac{E}{2R}\left(e^{-\frac{Rt}{L-M}} - e^{-\frac{Rt}{L+M}}\right). \qquad (32A.4)$$

NOTE: In a more comprehensive table of Laplace Transform pairs you would be able to find $\mathcal{L}^{-1}\left\{\frac{s+\alpha}{s(s+\beta)(s+\gamma)}\right\}$, *so that* i_1 *could be obtained from* $\bar{i}_1$ *without going through the stage of splitting into partial fractions.*

FRAME 33

For those who have studied both the D-operator and Laplace Transform methods, it is worth mentioning that the latter has certain advantages. In the L.T. method there is no need to consider the number of arbitrary constants required in the solution of simultaneous d.e.'s. Also, the processes involved in obtaining the solution of d.e.'s are of a routine nature due to the availability of comprehensive tables of transform pairs.

FRAME 34

Miscellaneous Examples

In this frame a collection of miscellaneous examples is given for you to try by either D-operator or Laplace Transform methods or both, unless otherwise stated. Answers are supplied in FRAME 35 and hints have been provided in some cases.

In examples 1 - 5 solve the given d.e.'s for x and y, with the initial conditions as stated.

1. $\frac{dx}{dt} + 4x + 3y = 0$

   $\frac{dy}{dt} + 3x + 4y = 2e^t$, given that $x = y = 0$ at $t = 0$.

FRAME 34 continued

2. $\quad 2\dfrac{dx}{dt} + \dfrac{dy}{dt} + 3y = 1$

$\quad \dfrac{dx}{dt} + \dfrac{dy}{dt} + y = e^{-t}$, given that $x = 0$, $y = 1$ at $t = 0$.

3. $\quad \dfrac{dx}{dt} + 7x - y = 0$

$\quad \dfrac{dy}{dt} + 2x + 5y = 0$, given that $x = 0$, $y = 1$ at $t = 0$.

4. $\quad Dx + (D - 2)y = 2\cos t - 7\sin t$

$\quad (D + 2)x - Dy = 4\cos t - 3\sin t$

$\quad$ where $D \equiv \dfrac{d}{dt}$, and $x = 3$, $y = 0$ at $t = 0$.

5. $\quad Dx + (D - 1)y = t$

$\quad (D + 1)x + 2Dy = e^t$

$\quad$ where $D \equiv \dfrac{d}{dt}$, and $x = 1$, $y = 0$ at $t = 0$.

6. Find the general solution of the following simultaneous d.e.'s by <u>the D-operator method</u>:

$\quad \dfrac{dx}{dt} + \dfrac{dy}{dt} + 2y = 1$

$\quad \dfrac{dx}{dt} + \dfrac{dy}{dt} + x = 2$.

7. Solve for y only, the following simultaneous d.e.'s <u>by the Laplace Transform method</u>:

$\quad (D - 2)x - (D - 2)y = \sin t$

$\quad (D^2 + 1)x + 2Dy = 0$

$\quad$ where $D \equiv \dfrac{d}{dt}$, and $x = Dx = y = 0$ at $t = 0$.

FRAME 34 continued

8. If an electron is projected into a uniform magnetic field which is perpendicular to its direction of motion, its path is given by

$$m\ddot{y} = -\frac{He}{c}\dot{x}$$

$$m\ddot{x} = \frac{He}{c}\dot{y}.$$

H, e, m and c are constants, and $x = 0$, $\dot{x} = u$, $y = \dot{y} = 0$ at $t = 0$. Find the coordinates of the position of the electron at time t.

(HINT: To simplify the algebra, put $k = \frac{He}{mc}$.)

9.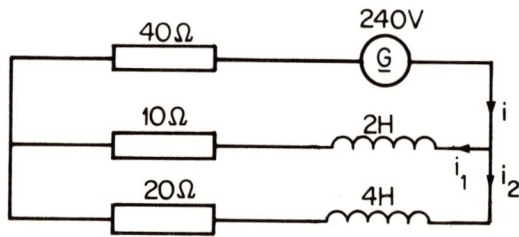

Find the currents in the three branches of this network, given that the initial currents are zero.

(If you are not familiar with electric networks, you will find the following hint for the derivation of the d.e.'s helpful.

Using Kirchhoff's Laws, we get

$$i = i_1 + i_2$$

$$40i - 240 + 2\frac{di_1}{dt} + 10i_1 = 0$$

$$-10i_1 - 2\frac{di_1}{dt} + 4\frac{di_2}{dt} + 20i_2 = 0.$$

Now, from the above equations, derive two simultaneous d.e.'s in i_1 and i_2.)

SOLUTION BY D-OPERATOR AND LAPLACE TRANSFORM METHODS

FRAME 34 continued

10. In the illustration in FRAME 2, if $R_1 = R_2 = 5 \,\Omega$, $L_1 = L_2 = 0.1$ H, $M = 0.05$ H, $e_1 = 10$ V and $e_2 = 0$, write down the two simultaneous d.e.'s for i_1 and i_2 and solve for the case when the currents are initially zero.

11. In the problem of the oscillation of two weights suspended one below the other by springs, as described in FRAME 1, put $m_1 = m_2 = 1$, $k_1 = 3$ and $k_2 = 2$ in equations (1.1) and show that, if the initial conditions are $x_1 = 5$, $x_2 = 0$, $\dot{x}_1 = \dot{x}_2 = 0$, the positions of the weights at time t are given by

$$x_1 = 4 \cos \sqrt{6}t + \cos t$$
$$x_2 = -2 \cos \sqrt{6}t + 2 \cos t.$$

FRAME 35

Answers to Miscellaneous Examples

1. $(D + 4)x + 3y = 0$
   $3x + (D + 4)y = 2e^t$

   By D-operator method:

   $\{(D + 4)^2 - 9\}x = -6e^{2t}$
   $(D + 7)(D + 1)x = -6e^{2t}$
   $x = Ae^{-7t} + Be^{-t} - \frac{2}{9}e^{2t}$.

   Substitution in first d.e. gives

   $y = Ae^{-7t} - Be^{-t} + \frac{4}{9}e^{2t}$.

   $x = y = 0$ when $t = 0$ gives $A = -\frac{1}{9}$, $B = \frac{1}{3}$.

   $\therefore x = -\frac{1}{9}e^{-7t} + \frac{1}{3}e^{-t} - \frac{2}{9}e^{2t}$

   $y = -\frac{1}{9}e^{-7t} - \frac{1}{3}e^{-t} + \frac{4}{9}e^{2t}.$

FRAME 35 continued

By Laplace Transform method:

$(s + 4)\bar{x} + 3\bar{y} = 0$

$3\bar{x} + (s + 4)\bar{y} = \dfrac{2}{s - 2}$

$\bar{x} = -\dfrac{6}{(s - 2)(s + 1)(s + 7)}$

$\phantom{\bar{x}} = -\dfrac{2}{9}\dfrac{1}{s - 2} + \dfrac{1}{3}\dfrac{1}{s + 1} - \dfrac{1}{9}\dfrac{1}{s + 7}$

$x = -\dfrac{2}{9}e^{2t} + \dfrac{1}{3}e^{-t} - \dfrac{1}{9}e^{-7t}$

y can then be found by substituting for x in the first d.e.

2.   By D-operator method:

Eliminating Dx gives $(D - 1)y = 2e^{-t} - 1$.
This is a linear first order d.e. and the integrating factor is e^{-t}.

$$y = -e^{-t} + 1 + Ae^t.$$

Substitution in the second d.e. gives $Dx = e^{-t} - 1 - 2Ae^t$

$x = -e^{-t} - t - 2Ae^t + B.$

The values of A and B are then found using the initial conditions.

$x = -e^{-t} - t - 2e^t + 3$

$y = -e^{-t} + e^t + 1.$

By Laplace Transform method:

$2s\bar{x} + (s + 3)\bar{y} = 1 + \dfrac{1}{s}$

$s\bar{x} + (s + 1)\bar{y} = 1 + \dfrac{1}{s + 1}$

$\bar{y} = \dfrac{1}{s - 1} + \dfrac{2}{(s - 1)(s + 1)} - \dfrac{1}{s(s - 1)}$

$y = e^t + (e^t - e^{-t}) + (1 - e^t)$

$ = 1 - e^{-t} + e^t$

SOLUTION BY D-OPERATOR AND LAPLACE TRANSFORM METHODS

FRAME 35 continued

$$\bar{x} = \frac{-2}{s(s-1)} - \frac{1}{s^2(s+1)}$$

$$= \frac{-2}{s(s-1)} + \frac{1}{s} - \frac{1}{s^2} - \frac{1}{s+1}$$

$$x = 2(1 - e^t) + 1 - t - e^{-t}$$

$$= 3 - 2e^t - t - e^{-t}.$$

3. By D-operator method:

Eliminating x gives $(D^2 + 12D + 37)y = 0$

$$y = e^{-6t}(A \cos t + B \sin t)$$

Substitution in second d.e. gives

$$x = \tfrac{1}{2}e^{-6t}\{(A - B) \cos t + (A + B) \sin t\}.$$

Initial conditions give $A = 1$, $B = 1$,

$$x = e^{-6t} \sin t$$

$$y = e^{-6t}(\cos t + \sin t).$$

By Laplace Transform method:

$(s + 7)\bar{x} - \bar{y} = 0$

$2\bar{x} + (s + 5)\bar{y} = 1$

$$\bar{x} = \frac{1}{(s+6)^2 + 1}$$

$$x = e^{-6t} \sin t$$

$$\bar{y} = \frac{s+7}{s^2 + 12s + 37} = \frac{s+6}{(s+6)^2 + 1} + \frac{1}{(s+6)^2 + 1}$$

$$y = e^{-6t} \cos t + e^{-6t} \sin t$$

$$= e^{-6t}(\cos t + \sin t).$$

FRAME 35 continued

4. By D-operator method:

Eliminating y gives $(D^2 - 2)x = -9 \cos t$

$$x = Ae^{\sqrt{2}t} + Be^{-\sqrt{2}t} + 3 \cos t.$$

Adding the two d.e.'s gives
$$(2D + 2)x - 2y = 6 \cos t - 10 \sin t.$$

Substituting for x, you will then get
$$y = (1 + \sqrt{2})Ae^{\sqrt{2}t} + (1 - \sqrt{2})Be^{-\sqrt{2}t} + 2 \sin t.$$

The initial conditions give $A = B = 0$

$$\therefore \quad x = 3 \cos t$$
$$y = 2 \sin t.$$

By Laplace Transform method:

$$s\bar{x} + (s - 2)\bar{y} = \frac{3s^2 + 2s - 4}{s^2 + 1}$$

$$(s + 2)\bar{x} - s\bar{y} = \frac{s(3s + 4)}{s^2 + 1}.$$

Solving for $\bar{x}$ and $\bar{y}$ leads to

$$\bar{x} = \frac{3s}{s^2 + 1} \quad \text{and} \quad \bar{y} = \frac{2}{s^2 + 1}$$

$$\therefore \quad x = 3 \cos t \qquad y = 2 \sin t.$$

5. By D-operator method:

Elimination gives $(D^2 + 1)y = e^t - t - 1$
$$y = A \cos t + B \sin t + \tfrac{1}{2}e^t - t - 1.$$

Subtracting the first d.e. from the second, you will get
$$x + Dy + y = e^t - t$$
from which $\quad x = (A - B) \sin t - (A + B) \cos t + 2.$

From the initial conditions, $A = B = \tfrac{1}{2}$.

$$\therefore \quad x = 2 - \cos t$$
$$y = \tfrac{1}{2} \cos t + \tfrac{1}{2} \sin t + \tfrac{1}{2}e^t - t - 1.$$

SOLUTION BY D-OPERATOR AND LAPLACE TRANSFORM METHODS 5:35

FRAME 35 continued

By Laplace Transform method:

$$s\bar{x} + (s - 1)\bar{y} = \frac{s^2 + 1}{s^2}$$

$$(s + 1)\bar{x} + 2s\bar{y} = \frac{s}{s - 1}.$$

Solving for $\bar{x}$ and $\bar{y}$ yields

$$\bar{y} = \frac{1}{s^2(s^2 + 1)(s - 1)} = -\frac{1}{s} - \frac{1}{s^2} + \frac{1}{2}\frac{s + 1}{s^2 + 1} + \frac{1}{2}\frac{1}{s - 1}$$

and $\bar{x} = \frac{s}{s^2 + 1} + \frac{2}{s(s^2 + 1)}$

$\therefore x = \cos t + 2(1 - \cos t) = 2 - \cos t$

and $y = -1 - t + \frac{1}{2}\cos t + \frac{1}{2}\sin t + \frac{1}{2}e^t$.

6. $x = Ae^{-2t/3} + 2$

$y = \frac{A}{2}e^{-2t/3} + \frac{1}{2}.$

7. $(s - 2)\bar{x} - (s - 2)\bar{y} = \frac{1}{s^2 + 1}$

$(s^2 + 1)\bar{x} + 2s\bar{y} = 0$

$\bar{y} = -\frac{1}{(s - 2)(s + 1)^2} = -\frac{1}{9}\frac{1}{s - 2} + \frac{1}{9}\frac{1}{s + 1} + \frac{1}{3}\frac{1}{(s + 1)^2}$

$y = -\frac{1}{9}e^{2t} + \frac{1}{9}e^{-t} + \frac{1}{3}te^{-t}.$

8. $x = \frac{u}{k}\sin kt$

$y = \frac{u}{k}(\cos kt - 1).$

FRAME 35 continued

9. The simultaneous d.e.'s in i_1 and i_2 are:

$$-(D + 5)i_1 + 2(D + 5)i_2 = 0$$

$$(D + 25)i_1 + 20i_2 = 120.$$

Solving these equations gives

$$i_1 = \frac{24}{7}(1 - e^{-35t}) \quad \text{and} \quad i_2 = \frac{12}{7}(1 - e^{-35t}).$$

The current in the third branch is

$$i = i_1 + i_2 = \frac{36}{7}(1 - e^{-35t}).$$

10. After re-arrangement the d.e.'s become

$$2(D + 50)i_1 + Di_2 = 200$$
$$Di_1 + 2(D + 50)i_2 = 0.$$

By D-operator method:

Eliminating i_1 gives $(3D + 100)(D + 100)i_2 = 0$

$$i_2 = Ae^{-100t/3} + Be^{-100t}.$$

Then eliminate Di_1 to obtain i_1 from i_2.

$$100i_1 - (3D + 200)i_2 = 200$$

whence $i_1 = 2 + Ae^{-100t/3} - Be^{-100t}$.

The conditions $i_1 = i_2 = 0$ at $t = 0$ give $A = -1$, $B = 1$.

By Laplace Transform method:

$$2(s + 50)\bar{i}_1 + s\bar{i}_2 = \frac{200}{s}$$

$$s\bar{i}_1 + 2(s + 50)\bar{i}_2 = 0$$

$$\bar{i}_2 = -\frac{200}{(s + 100)(3s + 100)}$$

$$i_2 = e^{-100t} - e^{-100t/3}$$

$$\bar{i}_1 = -\frac{2(s + 50)\bar{i}_2}{s} = \frac{400(s + 50)}{s(s + 100)(3s + 100)} = \frac{2}{s} - \frac{1}{s + 100} - \frac{3}{3s + 100}$$

$$i_1 = 2 - e^{-100t} - e^{-100t/3}.$$

FRAME 35 continued

11. After re-arrangement the d.e.'s become

$$(D^2 + 5)x_1 - 2x_2 = 0$$
$$-2x_1 + (D^2 + 2)x_2 = 0.$$

By D-operator method:

Either x_1 or x_2 can be eliminated with equal ease. Eliminating x_2 would give

$$(D^2 + 6)(D^2 + 1)x_1 = 0$$

whence $x_1 = A \cos \sqrt{6}t + B \sin \sqrt{6}t + C \cos t + E \sin t$.

Then substitute in the first d.e. to find x_2.

$$x_2 = -\frac{A}{2} \cos \sqrt{6}t - \frac{B}{2} \sin \sqrt{6}t + 2C \cos t + 2E \sin t.$$

Using the initial conditions you will find that
$A = 4$, $B = 0$, $C = 1$, $E = 0$.

By Laplace Transform method:

$$(s^2 + 5)\bar{x}_1 - 2\bar{x}_2 = 5s$$
$$-2\bar{x}_1 + (s^2 + 2)\bar{x}_2 = 0$$

$$\bar{x}_1 = \frac{5s(s^2 + 2)}{(s^2 + 6)(s^2 + 1)} = \frac{4s}{s^2 + 6} + \frac{s}{s^2 + 1}$$

$$\bar{x}_2 = \frac{2}{s^2 + 2}\bar{x}_1 = \frac{10s}{(s^2 + 6)(s^2 + 1)} = -\frac{2s}{s^2 + 6} + \frac{2s}{s^2 + 1}$$

APPENDIX

$$F(s) = \mathcal{L}\{f(t)\} = \int_0^\infty e^{-st} f(t)\, dt$$

TABLE OF LAPLACE TRANSFORM PAIRS

$f(t)$	$F(s)$
0	0
1	$1/s$
k	k/s
t	$1/s^2$
t^n	$n!/s^{n+1}$
$e^{\alpha t}$	$1/(s - \alpha)$
$\sin \omega t$	$\omega/(s^2 + \omega^2)$
$\cos \omega t$	$s/(s^2 + \omega^2)$
$\sin(\omega t + \phi)$	$(s \sin \phi + \omega \cos \phi)/(s^2 + \omega^2)$
$\cos(\omega t + \phi)$	$(s \cos \phi - \omega \sin \phi)/(s^2 + \omega^2)$
$\sinh \beta t$	$\beta/(s^2 - \beta^2)$
$\cosh \beta t$	$s/(s^2 - \beta^2)$
$\frac{1}{a-b}(e^{at} - e^{bt})$	$1/(s-a)(s-b)$
$\frac{1}{a-b}(ae^{at} - be^{bt})$	$s/(s-a)(s-b)$
$1 - e^{\alpha t}$	$-\alpha/s(s - \alpha)$
$1 - \cos \omega t$	$\omega^2/s(s^2 + \omega^2)$
$e^{\alpha t} f(t)$	$F(s - \alpha)$
$e^{\alpha t} t^n$	$n!/(s - \alpha)^{n+1}$
$e^{\alpha t} \sin \omega t$	$\omega/\{(s - \alpha)^2 + \omega^2\}$

SOLUTION BY D-OPERATOR AND LAPLACE TRANSFORM METHODS

APPENDIX

TABLE OF LAPLACE TRANSFORM PAIRS - continued

$f(t)$	$F(s)$
$e^{\alpha t} \cos \omega t$	$(s - \alpha)/\{(s - \alpha)^2 + \omega^2\}$
$t \sin \omega t$	$2\omega s/(s^2 + \omega^2)^2$
$t \cos \omega t$	$(s^2 - \omega^2)/(s^2 + \omega^2)^2$
$\sin \omega t - \omega t \cos \omega t$	$2\omega^3/(s^2 + \omega^2)^2$
$\dfrac{dx}{dt}$	$s\bar{x} - x_o$
$\dfrac{d^2 x}{dt^2}$	$s^2\bar{x} - sx_o - x_1$
$\dfrac{d^n x}{dt^n}$	$s^n\bar{x} - s^{n-1}x_o - s^{n-2}x_1 - \ldots - x_{n-1}$

LINEAR DIFFERENTIAL EQUATIONS

– VARIATION OF PARAMETERS and SOLUTION IN SERIES

A PROGRAMMED TEXT

A. C. Bajpai
I. M. Calus

INSTRUCTIONS

This programme constitutes a self-instructional course on two methods of solution of differential equations. The first of these is the method of variation of parameters and the second is the method of solution in series. The two sections are independent of one another and consequently either or both of them can be studied.

The programme is divided up into a number of FRAMES which are to be worked *in the order given*. You will be required to participate in many of these frames and in such cases the answers are provided in ANSWER FRAMES, designated by the letter A following the frame number. Steps in the working are given where this is considered helpful. The answer frame is separated from the main frame by a line of asterisks: ******************. Keep the answers covered until you have written your own response. If your answer is wrong, go back and try to see why. Do not proceed to the next frame until you have corrected any mistakes in your attempt and are satisfied that you understand the contents up to this point.

Before reading this programme it is necessary that you are familiar with the following

Prerequisites

For the Variation of parameters section:

    The contents of Programme 1, FRAMES 1 – 38 and 50 – 63.
    The contents of Programme 2, FRAMES 1 – 28.

For the Solution in Series section:

    No special reguirements for the main programme.
    Partial Differential coefficients for APPENDICES A and B.

CONTENTS

Instructions

VARIATION OF PARAMETERS

FRAMES

1	Introduction
2	Reduced Equation
3 - 8	First Order Linear Equations
9 - 19	Second Order Linear Equations
20	Miscellaneous Examples
21	Answers to Miscellaneous Examples

SOLUTION IN SERIES

FRAMES

1	Introduction
2	Method of Frobenius
3 - 8 } 13 - 21	Examples where roots of indicial equation do not differ by an integer
9 - 12	Recurrence Relation
22 - 34	Examples where roots of indicial equation differ by an integer
35	Roots of indicial equation equal
36 - 40	Legendre's Equation
41	Miscellaneous Examples
42	Answers to Miscellaneous Examples

APPENDICES

A	Roots of Indicial Equation equal
B	Roots of Indicial Equation differing by an integer and making an 'a' infinite
C	An example using $\sum$ notation

VARIATION OF PARAMETERS

VARIATION OF PARAMETERS

FRAME 1

Introduction

In the differential equations

$$y' - 2y = e^{3x} \qquad (1.1)$$

$$y' - \frac{3}{x}y = x^2 \qquad (1.2)$$

$$y'' - 5y' + 4y = 2e^{2x} \qquad (1.3)$$

$$y'' + y = \operatorname{cosec} x \qquad (1.4)$$

$$x^2 y'' + xy' - y = x^2 e^x \qquad (1.5)$$

the dependent function y and its derivatives all appear to the first degree only, i.e. the d.e.'s are linear.

Equations (1.1) and (1.2) are of the form $\frac{dy}{dx} + Py = Q$, and can be solved by using an integrating factor (I.F.) $e^{\int P dx}$, as shown in the programme "First Order Differential Equations" in this series. Equation (1.3) is of the form $ay'' + by' + cy = Q(x)$, i.e. a second order d.e. with constant coefficients, and various methods (trial solution, D-operator, Laplace Transform) for the solution of such equations have also been given in previous programmes. Although (1.4) is of the same type as (1.3), the function which appears on the R.H.S. is not one of the standard ones considered in the use of these methods. Equation (1.5) differs from (1.3) and (1.4) in that it does not have constant coefficients, and therefore cannot be solved by any of the methods so far considered.

This programme describes a technique which can be used in the solution of <u>all</u> the above equations, as well as in various other cases not considered here, such as problems involving systems of d.e.'s. Our aim is to give you a general appreciation of the concept involved in this technique by showing its application in a variety of situations, without necessarily implying that it should be used in preference to other methods where available.

FRAME 2

Reduced Equation

You will remember that the complementary function (C.F.) is that part of the general solution of
$$ay'' + by' + cy = Q(x) \qquad (2.1)$$
which is obtained by solving
$$ay'' + by' + cy = 0. \qquad (2.2)$$
The equation (2.2) is called the REDUCED EQUATION.

Note that:
- (i) In (2.1) all terms involving y appear on the L.H.S.
- (ii) In (2.2) the R.H.S. of (2.1), which consists of terms in x only, has been replaced by zero.

No doubt you can write down the reduced equations corresponding to equations (1.1) to (1.5).

****

2A

$y' - 2y = 0$ *is the reduced equation for (1.1)*

$y' - \dfrac{3}{x} y = 0$ " " " " " (1.2)

$y'' - 5y' + 4y = 0$ " " " " " (1.3)

$y'' + y = 0$ " " " " " (1.4)

$x^2 y'' + xy' - y = 0$ " " " " " (1.5)

FRAME 3

First Order Linear Equations

Let us consider first the d.e. (1.1), i.e.
$$y' - 2y = e^{3x}$$
whose reduced equation is
$$y' - 2y = 0.$$

VARIATION OF PARAMETERS

FRAME 3 continued

The general solution of this reduced equation, which is easily found by separating the variables, is $y = Ae^{2x}$.

To obtain the solution of the d.e. (1.1) we now set $y = ue^{2x}$, where u is a function of x, and then find what this function of x must be in order for (1.1) to be satisfied.

You will notice that in doing this, we are replacing the arbitrary constant A, i.e. the PARAMETER in the solution of the reduced equation, by a function u which varies with x. This procedure is therefore called the METHOD OF VARIATION OF PARAMETER.

FRAME 4

You may be wondering whether our assumption of a solution of the form $y = ue^{2x}$ is a justifiable one. In this example it is easy to show that it is, because you already know an alternative method for solving the equation. Multiplying throughout by the integrating factor e^{-2x}, the d.e. (1.1) can be rewritten

$$\frac{d}{dx}(e^{-2x}y) = e^{3x} \cdot e^{-2x}$$

On integration, $e^{-2x}y = e^{x} + c$

and so $y = e^{2x}(e^{x} + c)$ i.e. it is of the form ue^{2x}.

In fact, it is easy to see that this form of solution can be assumed for any d.e. whose reduced equation is $y' - 2y = 0$.

Such a d.e. could be written $y' - 2y = Q(x)$.

Multiplying through by the I.F. e^{-2x} yields

$$\frac{d}{dx}(e^{-2x}y) = Q(x)e^{-2x}$$

$$e^{-2x}y = \int Q(x)e^{-2x}dx + c$$

$$y = e^{2x}\left[\int Q(x)e^{-2x}dx + c\right] \text{ which is of the form } ue^{2x}.$$

FRAME 5

Having, we hope, set your mind at rest on this point, let us return to the problem of finding the function u for which $y = ue^{2x}$ satisfies the d.e. $y' - 2y = e^{3x}$.

If $y = ue^{2x}$, $y' = 2ue^{2x} + u'e^{2x}$ so we must have

$$2ue^{2x} + u'e^{2x} - 2ue^{2x} = e^{3x} \qquad (5.1)$$
$$u' = e^x$$
$$u = e^x + c$$
$$\therefore y = e^{2x}(e^x + c)$$

FRAME 6

Let us now apply the method of variation of parameter to the d.e. (1.2)

$$\text{i.e} \quad \frac{dy}{dx} - \frac{3}{x}y = x^2$$

whose reduced equation is $\frac{dy}{dx} - \frac{3}{x}y = 0$.

First, find the general solution of the reduced equation by separating the variables.

****

6A

$$\frac{dy}{y} = 3\frac{dx}{x}$$
$$y = cx^3$$

FRAME 7

Now set $y = ux^3$, and hence find the required solution.

****

VARIATION OF PARAMETERS

7A

$$3ux^2 + u'x^3 - \frac{3}{x}ux^3 = x^2 \qquad (7A.1)$$

$$u' = \frac{1}{x}$$

$$u = \log_e x + c$$

Notice that in equation (7A.1) the terms in u cancel out, as also happened in equation (5.1) in the previous example. Only u' remains, so that u can then be found by direct integration. This will, in fact, always happen when the trial solution of a linear first order d.e. is obtained by varying the parameter in the solution of the reduced equation.

FRAME 8

Solve the equation $L\frac{di}{dt} + Ri = E \sin \omega t$ by the method of variation of parameter. (This equation arises when an e.m.f. $E \sin \omega t$ is applied to a circuit consisting of an inductance L and a resistance R in series.)

****

8A

Reduced equation is $L\frac{di}{dt} + Ri = 0$
Solution of reduced equation is $i = Ae^{-Rt/L}$
Replacing A by a function of t gives

$$i = ue^{-Rt/L}$$

$$L\left(-\frac{R}{L}ue^{-Rt/L} + \frac{du}{dt}e^{-Rt/L}\right) + Rue^{-Rt/L} = E \sin \omega t$$

The terms in u cancel out, as they should do.

$$\frac{du}{dt} = \frac{E}{L}e^{Rt/L}\sin \omega t$$

$$u = \frac{E}{L}\frac{e^{Rt/L}}{\frac{R^2}{L^2} + \omega^2}\left(\frac{R}{L}\sin \omega t - \omega \cos \omega t\right) + c$$

$$= \frac{Ee^{Rt/L}}{R^2 + \omega^2 L^2}(R \sin \omega t - \omega L \cos \omega t) + c$$

$$i = \frac{E}{R^2 + \omega^2 L^2}(R \sin \omega t - \omega L \cos \omega t) + ce^{-Rt/L}$$

FRAME 9

Second Order Linear Equations

We shall now see how a similar technique can be applied to a second order d.e. such as (1.3), i.e.
$$y'' - 5y' + 4y = 2e^{2x}.$$

The reduced equation is $y'' - 5y' + 4y = 0$ and the solution of this (i.e. the C.F.) is $y = Ae^x + Be^{4x}$.

This has two parameters A and B. Replacing each of them by variables, u and v, say, we have
$$y = ue^x + ve^{4x}.$$

As with the first order equation considered in FRAMES 3 - 5, you already know another way of obtaining the solution. The C.F. has already been stated and the particular integral (P.I.) is easily found (by either trial solution or D-operator) to be $-e^{2x}$. Hence the general solution is
$$y = Ae^x + Be^{4x} - e^{2x}.$$

This can be written in a variety of ways, of which a few are
$$y = (A + e^x)e^x + (B - 2e^{-2x})e^{4x}$$
$$y = Ae^x + (B - e^{-2x})e^{4x}$$
$$y = (A - e^x)e^x + Be^{4x}$$

Each of these is of the form $y = ue^x + ve^{4x}$.

Again you can see that the assumed form of the solution is reasonable. You will, incidentally, notice that there are various possibilities for the functions u and v. Just three have been listed here; no doubt you can invent some more yourself. To reach the final solution it is sufficient to obtain just one of the various combinations of u and v.

VARIATION OF PARAMETERS   6:7

FRAME 10

Let us now return to the solution of the d.e. (1.3) by the method of variation of parameters. (Parameter$\underline{s}$ this time, as we have two of them.)

Having put $y = ue^x + ve^{4x}$
we then have $y' = ue^x + 4ve^{4x} + u'e^x + v'e^{4x}$.

It has already been shown that there are endless possibilities for u and v, any of which will give the required solution, so we choose those which can be obtained most easily.

As there are two functions u and v to be determined, two equations will be required. Substituting $ue^x + ve^{4x}$ for y in the original d.e. (1.3) will give one of these equations, but only one. The second equation is chosen to be

$$u'e^x + v'e^{4x} = 0 \qquad (10.1)$$

because this leads to the simplest possible working, as will be seen when this example is being completed.

Write down the expression for y", when the above condition is imposed.

****

10A

$$y'' = ue^x + 16ve^{4x} + u'e^x + 4v'e^{4x}$$

FRAME 11

We now have $y = ue^x + ve^{4x}$

$y' = ue^x + 4ve^{4x}$

$y'' = ue^x + 16ve^{4x} + u'e^x + 4v'e^{4x}$

Substitution in $y'' - 5y' + 4y = 2e^{2x}$ leads to

$$u'e^x + 4v'e^{4x} = 2e^{2x} \qquad (11.1)$$

You will notice that the u and v terms have cancelled out – this is because e^x and e^{4x} are solutions of the reduced equation.

FRAME 11 continued

We now have two equations (10.1) and (11.1) which have to be solved to obtain u and v. They involve u' and v' only, and can therefore be solved by simple algebra for u' and v', from which u and v can be found by integration. This straightforward process would not have been possible if any other equation had been chosen for (10.1).

Eliminating u' by subtraction gives

$$3v'e^{4x} = 2e^{2x}$$
$$v' = \frac{2}{3}e^{-2x}$$
$$v = c_1 - \frac{1}{3}e^{-2x}$$

Now find u, and hence write down the required solution.

****

11A

$$u'e^x = -\frac{2}{3}e^{-2x}e^{4x}$$
$$u' = -\frac{2}{3}e^x$$
$$u = c_2 - \frac{2}{3}e^x$$
$$y = (c_2 - \frac{2}{3}e^x)e^x + (c_1 - \frac{1}{3}e^{-2x})e^{4x}$$
$$= c_2 e^x + c_1 e^{4x} - e^{2x}$$

You will notice that the terms involving the arbitrary constants constitute the C.F. and the other part is the P.I.

FRAME 12

As has already been pointed out, the d.e. which we have just solved using variation of parameters could have equally well been solved by any of the methods described in previous programmes. This is because the exponential function on the R.H.S. is one which was covered by each of these methods.

VARIATION OF PARAMETERS

FRAME 12 continued

However, turning now to equation (1.4), i.e.
$$\frac{d^2y}{dx^2} + y = \operatorname{cosec} x$$
you will find yourself at a loss if, for instance, you try to find the P.I. by trial solution or D-operator. This is where the method of variation of parameters comes into its own.

The reduced equation is $\frac{d^2y}{dx^2} + y = 0$ and the solution of this is $y = A \cos x + B \sin x$.

Replacing the parameters A and B by variables u and v, we have

$$y = u \cos x + v \sin x$$
$$y' = -u \sin x + v \cos x + \boxed{u' \cos x + v' \sin x}$$

As before, the expression involving u' and v' (shown in the dotted rectangle) is set equal to zero, i.e.

$$u' \cos x + v' \sin x = 0 \qquad (12.1)$$

Then $\quad y'' = -u \cos x - v \sin x - u' \sin x + v' \cos x$

Substitution in $y'' + y = \operatorname{cosec} x$ leads to

$$-u' \sin x + v' \cos x = \operatorname{cosec} x \qquad (12.2)$$

Notice that, as in the example in FRAMES 10 and 11, we have two equations in u' and v', without any u or v terms.

Now solve (12.1) and (12.2) for u' and v', find u and v, and hence write down the required solution.

****

12A

$$u' = -1$$
$$u = -x + c_1$$
$$v' = \frac{\cos x}{\sin x}$$
$$v = \log_e \sin x + c_2$$

12A continued

Required solution is $y = (c_1 - x)\cos x + (c_2 + \log_e \sin x)\sin x$
$= c_1 \cos x + c_2 \sin x - x \cos x + \sin x \log_e \sin x$

FRAME 13

The d.e. $y'' - y = \dfrac{2}{1 + e^x}$, like the previous example, has a function on the R.H.S. which would present difficulties if you tried to use, for instance, trial solution or D-operator methods. You should now attempt its solution by the method of variation of parameters.

****

13A

Solution of reduced equation is $y = Ae^x + Be^{-x}$.

Set $y = ue^x + ve^{-x}$

$y' = ue^x - ve^{-x} + \boxed{u'e^x + v'e^{-x}}$

$\phantom{y' = ue^x - ve^{-x} + }$ Put $u'e^x + v'e^{-x} = 0$ \hfill (13A.1)

Then $y'' = ue^x + ve^{-x} + u'e^x - v'e^{-x}$

Substitution in the d.e. leads to

$$u'e^x - v'e^{-x} = \frac{2}{1 + e^x} \qquad (13A.2)$$

From (13A.1) and (13A.2), $\quad 2v'e^{-x} = -\dfrac{2}{1 + e^x}$

$$v' = -\frac{e^x}{1 + e^x}$$

$$v = -\log_e(1 + e^x) + c_1$$

$$u' = \frac{1}{e^x(1 + e^x)}$$

$$ = \frac{1}{e^x} - \frac{1}{e^x + 1} \quad as \quad \frac{1}{z(z+1)} = \frac{1}{z} - \frac{1}{z+1}$$

$$ = e^{-x} - \frac{e^{-x}}{1 + e^{-x}}$$

VARIATION OF PARAMETERS

13A continued

$$u = -e^{-x} + \log_e(1 + e^{-x}) + c_2$$

Required solution is $y = e^x\left[c_2 - e^{-x} + \log_e(1 + e^{-x})\right] + e^{-x}\left[c_1 - \log_e(1 + e^x)\right]$

FRAME 14

In the three second order differential equations which we have just solved the coefficients of y, $\frac{dy}{dx}$ and $\frac{d^2y}{dx^2}$ were constant, and we were therefore able to make use of the forms of the C.F. already known for such types of equation. These forms will not apply when the coefficients of y and its derivatives are not constant, as in equation (1.5). In such cases, therefore, the first problem is to find the solution of the reduced equation, and this may be difficult. Certain rules of guidance on suitable trial solutions can be given. However, this problem will not be dealt with here as the main concern of this programme is to show the application of the technique of variation of parameters. We shall therefore start from the point where the C.F. (or part of it) has been found.

FRAME 15

In the case of equation (1.5), i.e.
$$x^2 y'' + xy' - y = x^2 e^x$$
the solutions $y = Ax$ and $y = \frac{B}{x}$ of the reduced equation can be found, giving the C.F. as $Ax + \frac{B}{x}$.

Using the method of variation of parameters to solve (1.5) we now put $y = ux + \frac{v}{x}$ and proceed in the usual way.

By this time, you should have no difficulty in writing down y', and then the first equation in u' and v'.

****

15A

$$y' = u - \frac{v}{x^2} + \boxed{u'x + \frac{v'}{x}}$$
$$u'x + \frac{v'}{x} = 0 \qquad (15A.1)$$

FRAME 16

Now go ahead and write down y'' and the second equation in u' and v'. Solve for u' and v', and hence find u and v so that you can write down the general solution of the d.e. (1.5).

****

16A

$$y'' = \frac{2v}{x^3} + u' - \frac{v'}{x^2}$$

Substitution in (1.5) leads to $x^2(u' - \frac{v'}{x^2}) = x^2 e^x$

$$u' - \frac{v'}{x^2} = e^x \qquad (16A.1)$$

Equations (15A.1) and (16A.1) give

$$u' = \tfrac{1}{2}e^x$$
$$v' = -\tfrac{1}{2}x^2 e^x$$

Hence $u = \tfrac{1}{2}e^x + c_1$ and $v = -\tfrac{1}{2}x^2 e^x + \int x e^x dx$

$$= -\tfrac{1}{2}x^2 e^x + x e^x - e^x + c_2$$

Required solution is $y = (\tfrac{1}{2}e^x + c_1)x + (-\tfrac{1}{2}x^2 e^x + x e^x - e^x + c_2)\tfrac{1}{x}$

$$= c_1 x + \frac{c_2}{x} + (1 - \frac{1}{x})e^x$$

FRAME 17

Reference has already been made to the difficulties which may occur in trying to find the C.F. of a second order d.e. with variable coefficients. However even if only part of it can be spotted, it is still possible to use the method of variation of parameters to find the general solution of the d.e.

For instance, in the case of the equation just solved, i.e.

$$x^2 y'' + xy' - y = x^2 e^x$$

it is easily verified that $y = x$, and hence $y = Ax$, is a solution of the reduced equation. Suppose that the solution $y = \frac{1}{x}$, and therefore $y = \frac{B}{x}$, had escaped notice.

VARIATION OF PARAMETERS

FRAME 17 continued

Proceeding on this basis, we vary the one parameter in the known part of the C.F. and put $y = ux$.

This gives $y' = u + u'x$
and $y'' = 2u' + u''x$.

What equation is obtained when these expressions are substituted in the d.e.?

****

17A

$$u''x + 3u' = e^x \qquad (17A.1)$$

FRAME 18

The substitution $p = u'$ in (17A.1) leads to a first order d.e. in p.

$$xp' + 3p = e^x$$

i.e. $$p' + \frac{3}{x}p = \frac{1}{x}e^x$$

Multiply throughout by the appropriate integrating factor and hence solve for p.

****

18A

$$I.F. = e^{3\log_e x} = x^3$$

$$\frac{d}{dx}(x^3 p) = x^2 e^x$$

$$x^3 p = x^2 e^x - \int 2x e^x dx$$

$$= x^2 e^x - 2x e^x + 2 e^x + c_1$$

$$p = \frac{1}{x} e^x - \frac{2}{x^2} e^x + \frac{2}{x^3} e^x + \frac{c_1}{x^3}$$

FRAME 19

We now have
$$u' = \frac{1}{x}e^x - \frac{2}{x^2}e^x + \frac{2}{x^3}e^x + \frac{c_1}{x^3}$$
$$= \left(\frac{1}{x} - \frac{1}{x^2}\right)e^x + \left(-\frac{1}{x^2} + \frac{2}{x^3}\right)e^x + \frac{c_1}{x^3}$$

Noting that $\left(\frac{1}{x} - \frac{1}{x^2}\right)e^x + \left(-\frac{1}{x^2} + \frac{2}{x^3}\right)e^x = \frac{d}{dx}\left\{\left(\frac{1}{x} - \frac{1}{x^2}\right)e^x\right\}$, we have

$$u = \left(\frac{1}{x} - \frac{1}{x^2}\right)e^x - \frac{c_1}{2x^2} + c_2$$

$$y = ux = \left(1 - \frac{1}{x}\right)e^x - \frac{c_1}{2x} + c_2 x$$

$$= c_2 x + \frac{C}{x} + \left(1 - \frac{1}{x}\right)e^x \qquad \text{where } C = -\frac{c_1}{2}$$

It is worth noting that, in general:

The d.e. in u (17A.1 in this case) is of the second order and does not contain u. It can therefore be reduced to a first order d.e. by the substitution $p = u'$.

FRAME 20

Miscellaneous Examples

In this frame, a collection of miscellaneous examples is given for you to try. Answers are supplied in FRAME 21 and hints have been provided in some cases.

Use the method of variation of parameters to solve the following equations.

1. $y' + y \cos x = \frac{1}{2} \sin 2x$

2. $y'' + n^2 y = \sec nx$

3. $y'' - 2y' + y = e^x$

4. Given that $y = x$ and $y = e^x$ are solutions of the equation
$$(x - 1)y'' - xy' + y = 0$$
obtain the solution of
$$(x - 1)y'' - xy' + y = (x - 1)^2$$

VARIATION OF PARAMETERS   6:15

                                                              FRAME 20 continued

5.   It is easy to see that $y = x$ is a solution of
     $$x^2 y'' - 2xy' + 2y = 0.$$
     Use this result to solve the equation
     $$x^2 y'' - 2xy' + 2y = x^3.$$

6.   Standard trial solution techniques yield the solution $y = e^x$ for the equation $xy'' - (x+1)y' + y = 0$. Use this to solve the equation $xy'' - (x+1)y' + y = 2x^2 e^x$.

                                                                     FRAME 21

Answers to Miscellaneous Examples

1.   Solution of reduced equation: $y = Ae^{-\sin x}$.

     Put $y = ue^{-\sin x}$

     $u = \int \sin x (e^{\sin x} \cos x) dx$, which can be integrated by parts.

     $y = c_1 e^{-\sin x} + \sin x - 1$

2.   C.F. is $A \cos nx + B \sin nx$.

     Put $y = u \cos nx + v \sin nx$.

     $\left. \begin{array}{l} u' \cos nx + v' \sin nx = 0 \\ -u'n \sin nx + v'n \cos nx = \sec nx \end{array} \right\}$ give $u' = -\frac{1}{n} \tan nx$ and $v' = \frac{1}{n}$

     $y = (c_1 + \frac{1}{n^2} \log_e \cos nx) \cos nx + (c_2 + \frac{x}{n}) \sin nx$

3.   C.F. is $Ae^x + Bxe^x$.

     Put $y = ue^x + vxe^x$.

     $\left. \begin{array}{l} u' + v'x = 0 \\ u' + v'(x+1) = 1 \end{array} \right\}$ give $u' = -x$ and $v' = 1$

     $y = (c_1 + c_2 x + \frac{x^2}{2}) e^x$

     NOTE: This would be a "case of failure" if trial solution or D-operator methods were used.

FRAME 21 continued

4. Put $y = ux + ve^x$

$$\left.\begin{array}{l} u'x + v'e^x = 0 \\ u' + v'e^x = x - 1 \end{array}\right\} \text{ give } u' = -1 \text{ and } v' = xe^{-x}$$

$$\begin{aligned} y &= (c_1 - x)x + (c_2 - xe^{-x} - e^{-x})e^x \\ &= c_1 x + c_2 e^x - x^2 - x - 1 \end{aligned}$$

5. Put $y = ux$

$u'' = 1$ (The d.e. in u is especially simple in this example because there is no term in u'.)

$$y = \left(\frac{x^2}{2} + c_1 x + c_2\right)x$$

6. Put $y = ue^x$

$$p' + \left(1 - \frac{1}{x}\right)p = 2x \quad \text{where} \quad p = u'$$

$$p = 2x + c_1 xe^{-x}$$

$$y = c_2 e^x - c_1(x + 1) + x^2 e^x$$

SOLUTION IN SERIES

SOLUTION IN SERIES

FRAME 1

Introduction

Previous programmes in this series have described methods for solving linear d.e.'s which are either (i) first order, or (ii) higher order with constant coefficients. Of these methods the only one that is applicable to linear d.e.'s of order greater than one, when the coefficients are not constant, is the method of variation of parameters, and even this is limited to cases where at least one solution of the reduced equation is known. A method which is more generally effective in the solution of linear d.e.'s with variable coefficients is based on the use of power series.

The idea of a function being expressed as a power series is no doubt familiar to you, and, for instance, it is the series form which a digital computer uses when calculating the value of a function. Many of the solutions of d.e.'s which you have found in functional form could have been expressed in a series form. Thus, for example, the equation $\dfrac{d^2y}{dx^2} + n^2 y = 0$ has the general solution

$$y = A \cos nx + B \sin nx$$

which could also be written as

$$y = A\left(1 - \frac{n^2 x^2}{2!} + \frac{n^4 x^4}{4!} - \ldots\right) + B\left(nx - \frac{n^3 x^3}{3!} + \frac{n^5 x^5}{5!} - \ldots\right)$$

Some differential equations which occur in science and engineering cannot be solved in terms of the standard functions known to us but yet one can find solutions for them in the form of an infinite series. In fact one can take the point of view that a differential equation is a way of defining a new function, values of which can be tabulated from the series for it. One such equation is Bessel's equation, which occurs in the theory of vibrations, heat flow and the propagation of electricity in conductors. The equation

$$x^2 \frac{d^2 y}{dx^2} + x \frac{dy}{dx} + (x^2 - n^2) y = 0$$

is called Bessel's equation of order n and its solutions are called Bessel functions of order n.

FRAME 2

Method of Frobenius

Sometimes a series solution to a d.e. can be found by assuming a Maclaurin expansion, sometimes a trial solution

$$y = a_0 + a_1 x + a_2 x^2 + \ldots + a_r x^r + \ldots$$

is successful. A more general method (the METHOD OF FROBENIUS) is to take as a trial solution

$$y = x^c (a_0 + a_1 x + a_2 x^2 + \ldots + a_r x^r + \ldots) \qquad (2.1)$$

where a_0 is the first non-zero coefficient in the expansion.

The existence of such a solution is dependent on certain properties of the coefficients in the d.e. Also, just as you will recall, the expansion of a function in a Maclaurin series is only valid if the series converges, so also is the series solution of a d.e. dependent on the convergence of the series, if infinite, and it may therefore only be valid over a limited range of values of x. Problems of determining whether a series solution exists and, if it does, for what range of values it is convergent, will not be dealt with in this programme. The main purpose is to show how, in equations which have solutions of the form (2.1), these solutions can be found.

FRAME 3

The value of c and the coefficients a_r have to be found. This is done by substituting the trial solution in the given d.e. and equating the coefficients of like powers of x in the resulting identity. A few examples will show you how the method works.

Example 1   Solve the equation

$$4xy'' + 2y' + y = 0 \qquad (3.1)$$

SOLUTION IN SERIES

FRAME 3 continued

The trial solution (2.1) gives

$$y = a_0 x^c + a_1 x^{c+1} + a_2 x^{c+2} + \ldots$$

$$y' = a_0 c x^{c-1} + a_1(c+1)x^c + a_2(c+2)x^{c+1} + \ldots$$

$$y'' = a_0 c(c-1)x^{c-2} + a_1(c+1)c x^{c-1} + a_2(c+2)(c+1)x^c + \ldots$$

(These equations are basic to the method of Frobenius, and will be required in every example. You may find it useful to copy them out, to keep by you for easy reference.)

When these expressions are substituted in (3.1), an identity with zero on the R.H.S. is obtained. Thus the coefficient of each power of x on the L.H.S. must be zero. We shall start with the coefficient of the <u>lowest</u> power of x, so the first question to be answered is "What is the lowest power of x obtained when the trial solution is substituted in $4xy'' + 2y' + y$?" Write down your answer to this question. Don't bother about the coefficient itself at this stage.

****

3A

x^{c-1}, *coming from the $4xy''$ and $2y'$ terms.*

FRAME 4

The next step is to write down the coefficient of x^{c-1} and equate it to zero. The equation thus obtained is

$$4a_0 c(c-1) + 2a_0 c = 0$$

$$a_0 c(2c - 1) = 0$$

$a_0 \neq 0$ (it was defined in FRAME 2 as the first non-zero coefficient)

$$\therefore c(2c - 1) = 0$$

This equation in c is called the INDICIAL EQUATION. Here there are two possibilities for c, $c = 0$ and $c = \frac{1}{2}$.

FRAME 5

Continuing in this way, equating the coefficients of x^c, x^{c+1}, etc. to zero, the a's can be found.

x^c gives
$$4a_1(c+1)c + 2a_1(c+1) + a_0 = 0$$
$$2a_1(c+1)(2c+1) + a_0 = 0$$
$$a_1 = -\frac{a_0}{2(c+1)(2c+1)} \qquad (5.1)$$

We now have a_1 in terms of a_0. If you equate the coefficient of x^{c+1} to zero, you should obtain a_2 in terms of a_1.

***

5A

$$4a_2(c+2)(c+1) + 2a_2(c+2) + a_1 = 0 \qquad (5A.1)$$
$$a_2 = -\frac{a_1}{2(c+2)(2c+3)} \qquad (5A.2)$$

FRAME 6

You could now continue this process, equating the coefficients of higher powers of x to zero, but by this time a pattern is emerging. If you can see it, you will be able to write down a_3 in terms of a_2, without reference to the series for y, y' and y" in FRAME 3. If you can't, proceed as before, this time equating the coefficient of x^{c+2} to zero and then see if you can write down a_4 in terms of a_3.

***

6A

$$a_3 = -\frac{a_2}{2(c+3)(2c+5)} \qquad (6A.1)$$

If you were not able to write this down straight away, you should have obtained the equation

$$4a_3(c+3)(c+2) + 2a_3(c+3) + a_2 = 0$$

SOLUTION IN SERIES   6:23

6A continued

Comparing this with 5A.1, you will see that $(c + 3)$ replaces $(c + 2)$ and $(c + 2)$ replaces $(c + 1)$, i.e. c has everywhere been replaced by $(c + 1)$. When this is done in (5A.2), the relationship (6A.1) is obtained, from which in turn

$$a_4 = -\frac{a_3}{2(c+4)(2c+7)}$$

can be derived in a similar way.

FRAME 7

Having established a relationship between successive coefficients we now find what series will result from each of the values of c obtained from the indicial equation in FRAME 4.

Taking first $c = 0$, and substituting in (5.1), (5A.1) and (6A.1), we have

$$a_1 = -\frac{a_0}{2}$$

$$a_2 = -\frac{a_1}{4 \cdot 3} = \frac{a_0}{4!}$$

$$a_3 = -\frac{a_2}{6 \cdot 5} = -\frac{a_0}{6!}$$

You will notice that a_1, a_2, a_3 etc. can all be expressed in terms of a_0, which itself remains undetermined. Thus $c = 0$ yields the solution

$$a_0 x^0 \left(1 - \frac{x}{2!} + \frac{x^2}{4!} - \frac{x^3}{6!} + \ldots\right)$$

But $x^0 = 1$, and as a_0 can have any value, we can represent it by an arbitrary constant A. The solution becomes

$$A\left(1 - \frac{x}{2!} + \frac{x^2}{4!} - \frac{x^3}{6!} + \ldots\right)$$

Now, taking $c = \frac{1}{2}$, write down a_1, a_2 and a_3 in terms of a_0 and hence obtain the solution

$$Bx^{\frac{1}{2}}\left(1 - \frac{x}{3!} + \frac{x^2}{5!} - \frac{x^3}{7!} + \ldots\right)$$

where B is another arbitrary constant.

****

7A

$$a_1 = -\frac{a_0}{3.2} = -\frac{a_0}{3!}$$

$$a_2 = -\frac{a_1}{5.4} = \frac{a_0}{5!}$$

$$a_3 = -\frac{a_2}{7.6} = -\frac{a_0}{7!}$$

Again, all coefficients are multiples of a_0, which is itself undetermined and can therefore be represented by an arbitrary constant B:

FRAME 8

We now have the two independent solutions

$$A\left(1 - \frac{x}{2!} + \frac{x^2}{4!} - \frac{x^3}{6!} + \ldots\right)$$

and $$Bx^{\frac{1}{2}}\left(1 - \frac{x}{3!} + \frac{x^2}{5!} - \frac{x^3}{7!} + \ldots\right)$$

and it can be shown that their sum is also a solution. (Your study of the equation $a\frac{d^2y}{dx^2} + b\frac{dy}{dx} + cy = 0$ will have familiarised you with this property of linear d.e.'s.) Furthermore, the sum of these two solutions will have the two arbitrary constants necessary to the general solution of a second order d.e.

$$\therefore y = A\left(1 - \frac{x}{2!} + \frac{x^2}{4!} - \frac{x^3}{6!} + \ldots\right) + Bx^{\frac{1}{2}}\left(1 - \frac{x}{3!} + \frac{x^2}{5!} - \frac{x^3}{7!} + \ldots\right)$$

is the general solution of the d.e. (3.1).

FRAME 9

Recurrence Relation

In Example 1 we equated to zero the coefficients of x^{c-1}, x^c, x^{c+1} etc. in the expression obtained when the trial solution was substituted in the L.H.S. of the d.e. We found that equating the coefficient of the lowest power of x (x^{c-1}) to zero gave the indicial equation from which values of the index c

SOLUTION IN SERIES

FRAME 9 continued

could be found. From x^c onwards, equating the coefficients to zero gave relationships between successive coefficients i.e. a_1 and a_0, a_2 and a_1, and so on. These were:

$$a_1 = -\frac{a_0}{2(c+1)(2c+1)}$$

$$a_2 = -\frac{a_1}{2(c+2)(2c+3)}$$

$$a_3 = -\frac{a_2}{2(c+3)(2c+5)}$$

$$a_4 = -\frac{a_3}{2(c+4)(2c+7)}$$

The relationships are following a set pattern and it should therefore be possible to write down a general formula connecting <u>any</u> two successive coefficients.

Try to write down a relationship giving a_{r+1} in terms of a_r. Check that it works for the particular coefficients listed above and then turn to the answer frame to be quite sure that your answer is correct.

****

9A

$$a_{r+1} = -\frac{a_r}{2(c+r+1)(2c+2r+1)}$$

FRAME 10

This equation connecting a_{r+1} and a_r is an example of a RECURRENCE RELATION (sometimes called RECURSION FORMULA). In this particular case, it holds for all $r \geq 0$.

In fact, the formula for a_{r+1} could have been written down in the first place, without writing down the formulae for a_1, a_2 etc.

Equating the coefficient of x^c to zero gave a_1 in terms of a_0.
" " " " x^{c+1} " " " a_2 " " " a_1.
" " " " x^{c+2} " " " a_3 " " " a_2.

FRAME 10 continued

If instead we had taken the coefficient of the general term x^{c+r} and equated it to zero, this would have led to a relation between which two coefficients?

****
10A

a_{r+1} in terms of a_r.

FRAME 11

To write down the coefficient of x^{c+r} when the trial solution is substituted in $4xy'' + 2y' + y$, we shall require:

the coefficient of x^{c+r-1} in y''
" " " x^{c+r} " y'
" " " x^{c+r} " y.

These terms are shown below.

$y = a_0 x^c + \ldots\ldots\ldots\ldots + a_r x^{c+r} + a_{r+1} x^{c+r+1} + .$

$y' = a_0 c x^{c-1} + \ldots\ldots\ldots\ldots + a_{r+1}(c+r+1)x^{c+r} + \ldots\ldots\ldots$

$y'' = a_0 c(c-1)x^{c-2} + \ldots\ldots + a_{r+1}(c+r+1)(c+r)x^{c+r-1} + \ldots\ldots\ldots\ldots$

With the necessary information laid out in this way you should find it a simple matter to write down the coefficient of x^{c+r} in $4xy'' + 2y' + y$.

****
11A

$4a_{r+1}(c+r+1)(c+r) + 2a_{r+1}(c+r+1) + a_r$

simplifying to $2a_{r+1}(c+r+1)(2c+2r+1) + a_r$

SOLUTION IN SERIES

FRAME 12

Equating this coefficient to zero gives the recurrence relation

$$a_{r+1} = -\frac{a_r}{2(c+r+1)(2c+2r+1)}$$

as previously arrived at in answer frame 9A. Although obtaining a recurrence relation from the coefficient of x^{c+r} is really the more efficient way of obtaining the a's for the series, you may feel it is easier to write down a_1, a_2, etc. as was done in Example 1. The important thing is that you understand what you are doing.

FRAME 13

The example which now follows is similar to Example 1 and should not present any difficulties when you are called upon to participate in solving it.

Example 2   Solve the equation

$$2x^2 y'' - xy' + (1 - 2x)y = 0 \qquad (13.1)$$

You first have to decide what will be the lowest power of x when the trial solution $y = x^c(a_0 + a_1 x + a_2 x^2 + ..)$ is substituted in the L.H.S. of (13.1). Having done that, equate its coefficient to zero, giving the indicial equation, and solve for c.

****

13A

Lowest power: x^c

$$2a_0 c(c-1) - a_0 c + a_0 = 0$$

$$(c-1)(2c-1) = 0$$

$$c = 1 \quad or \quad c = \tfrac{1}{2}$$

FRAME 14

Now find the series solution corresponding to each value of c (it will be sufficient to write down the first four terms) and hence the general solution of the d.e. (13.1).

****

14A

The coefficients of x^{c+1} onwards all follow the same pattern. You could have equated the coefficient of x^{c+1} to zero, giving

$$2a_1(c + 1)c - a_1(c + 1) + a_1 - 2a_0 = 0$$

whence $a_1 = \dfrac{2a_0}{c(2c + 1)}$

and it could then be seen that $a_2 = \dfrac{2a_1}{(c + 1)(2c + 3)}$, $a_3 = \dfrac{2a_2}{(c + 2)(2c + 5)}$

etc. Alternatively, you could have equated the coefficient of x^{c+r} to zero, giving $2a_r(c + r)(c + r - 1) - a_r(c + r) + a_r - 2a_{r-1} = 0$

$$a_r = \dfrac{2a_{r-1}}{(c + r - 1)(2c + 2r - 1)} \quad \text{for} \quad r \geq 1$$

This time the coefficient of x^{c+r} gives a_r in terms of a_{r-1}, instead of a_{r+1} in terms of a_r, but you will realise that this does not matter — the essential feature of the recurrence relation in these examples is that it gives one 'a' in terms of a previous one.

$c = 1$ gives $a_1 = \dfrac{2a_0}{1.3}$ $a_2 = \dfrac{2a_1}{2.5}$ $a_3 = \dfrac{2a_2}{3.7}$

Denoting a_0 by an arbitrary constant A, this gives the solution

$$Ax\left[1 + \dfrac{2}{1!3}x + \dfrac{2^2}{2!3.5}x^2 + \dfrac{2^3}{3!3.5.7}x^3 + \ldots\right] \quad (14A.1)$$

$c = \tfrac{1}{2}$ gives $a_1 = \dfrac{2a_0}{1.1}$ $a_2 = \dfrac{2a_1}{3.2}$ $a_3 = \dfrac{2a_2}{5.3}$

Denoting a_0 by an arbitrary constant B, this solution is

$$Bx^{\tfrac{1}{2}}\left[1 + 2x + \dfrac{2^2}{2!3}x^2 + \dfrac{2^3}{3!3.5}x^3 + \ldots\right] \quad (14A.2)$$

The general solution is the sum of (14A.1) and (14A.2).

SOLUTION IN SERIES   6:29

FRAME 15

In both Examples 1 and 2, the recurrence relation was between a coefficient and the one before it, i.e. between a_{r+1} and a_r or a_r and a_{r-1}. This is not necessarily always the case, and in the next example you will find that alternate coefficients are related, i.e. a_r and a_{r-2}.

Example 3   Solve the equation

$$x^2 y'' + xy' + (x^2 - \tfrac{1}{9})y = 0$$

Taking the usual trial solution $y = x^c(a_0 + a_1 x + a_2 x^2 + \ldots)$ obtain the indicial equation and solve for c.

(If you have found difficulty in writing down the coefficients of the various powers of x using only the series for y, y' and y", as displayed in FRAMES 3 and 11, you may find the lay-out shown on Page 6:30 helpful.)

****

15A

$$a_0 c(c-1) + a_0 c - \tfrac{1}{9} a_0 = 0$$

$$c^2 - \tfrac{1}{9} = 0$$

$$c = \pm \tfrac{1}{3}$$

FRAME 16

Your experience of the series solution method up to this point will have shown you that, when the trial solution is substituted in the L.H.S. of the d.e., the only 'a' involved in the coefficient of the lowest power of x is a_0. In Examples 1 and 2 the coefficients of all powers of x after that satisfied the recurrence relation involving two 'a's.

In the present example, the indicial equation was obtained from the coefficient of x^c. Now write down the coefficient of the next higher power of x, i.e. x^{c+1}, and see how the situation this time differs from that in the previous examples.

****

Suggested lay-out (see FRAME 15) for obtaining the coefficients of powers of x required in the solution of the equation $x^2 y'' + xy' + (x^2 - \frac{1}{9})y = 0$

The following are written down using the expressions for y, y' and y'' given in FRAME 3:

$x^2 y'' = a_o c(c-1)x^c + a_1(c+1)cx^{c+1} + a_2(c+2)(c+1)x^{c+2} + \ldots\ldots + a_r(c+r)(c+r-1)x^{c+r} + \ldots\ldots$

$xy' = a_o cx^c + a_1(c+1)x^{c+1} + a_2(c+2)x^{c+2} + \ldots\ldots + a_r(c+r)x^{c+r} + \ldots\ldots$

$x^2 y = a_o x^{c+2} + \ldots\ldots + a_{r-2} x^{c+r} + \ldots\ldots$

$-\frac{1}{9}y = -\frac{1}{9}a_o x^c - \frac{1}{9}a_1 x^{c+1} - \frac{1}{9}a_2 x^{c+2} - \ldots\ldots - \frac{1}{9}a_r x^{c+r} - \ldots\ldots$

Adding vertically:

Coefficient of x^c = $a_o c(c-1) + a_o c - \frac{1}{9}a_o$

" x^{c+1} = $a_1(c+1)c + a_1(c+1) - \frac{1}{9}a_1$

" x^{c+2} = $a_2(c+2)(c+1) + a_2(c+2) + a_o - \frac{1}{9}a_2$

and so on.

SOLUTION IN SERIES

16A

$$a_1(c+1)c + a_1(c+1) - \frac{1}{9}a_1$$

This involves only a_1, not a_1 and a_0.

FRAME 17

Equating the coefficient of x^{c+1} to zero gives

$$a_1\{(c+1)^2 - \frac{1}{9}\} = 0 \qquad (17.1)$$

Neither of the values of c obtained from the indicial equation will make the expression in the curly brackets zero, so we must have $a_1 = 0$.

If you now write down the coefficient of x^{c+2} and equate it to zero you will find you have a relation between a_2 and a_0. Try to see what difference between the coefficient of x^{c+2} and those of x^c and x^{c+1} has caused two 'a's to be involved instead of only one.

****

17A

$$a_2(c+2)(c+1) + a_2(c+2) + a_0 - \frac{1}{9}a_2 = 0 \qquad (17A.1)$$

<u>All</u> terms in $x^2y'' + xy' + (x^2 - \frac{1}{9})y = 0$ are now contributing. The x^2y term did not make a contribution before.

(17A.1) may be written

$$a_2\{(c+2)^2 - \frac{1}{9}\} + a_0 = 0$$

giving

$$a_2 = -\frac{9a_0}{(3c+5)(3c+7)}$$

FRAME 18

Equating the coefficient of x^{c+r} to zero will give a_r in terms of a_{r-2}. Find this recurrence relation and say for what values of r it is valid.

****

18A

$$a_r\{(c + r)^2 - \tfrac{1}{9}\} + a_{r-2} = 0$$

$$a_r = -\frac{9a_{r-2}}{(3c + 3r - 1)(3c + 3r + 1)} \qquad \text{Valid for } r \geq 2$$

FRAME 19

What can you now say about a_3, a_5, a_7 etc. when c is either $+\tfrac{1}{3}$ or $-\tfrac{1}{3}$?

****

19A

They are all zero, because $a_1 = 0$ and a_3, a_5, a_7 etc. are all multiples of a_1.

FRAME 20

Now put $c = \tfrac{1}{3}$ in the recurrence relation given in 18A. Use the result to write down a_2, a_4 and a_6, and hence the first four terms in the series solution corresponding to this value of c.

****

20A

$$a_r = -\frac{3a_{r-2}}{r(3r + 2)}$$

$$a_2 = -\frac{3}{2.8} a_0$$

$$a_4 = -\frac{3}{4.14} a_2 = \frac{3^2}{2.4.8.14} a_0$$

$$a_6 = -\frac{3}{6.20} a_4 = -\frac{3^3}{2.4.6.8.14.20} a_0$$

$$Ax^{1/3}\left(1 - \frac{3}{2.8} x^2 + \frac{3^2}{2.4.8.14} x^4 - \frac{3^3}{2.4.6.8.14.20} x^6 \ldots \right)$$

SOLUTION IN SERIES

FRAME 21

You can now complete the solution to this problem by finding the series corresponding to $c = -\frac{1}{3}$.

***

21A

$$a_r = -\frac{3a_{r-2}}{r(3r-2)}$$

$$Bx^{-1/3}\left\{1 - \frac{3}{2.4}x^2 + \frac{3^2}{2.4.4.10}x^4 - \frac{3^3}{2.4.6.4.10.16}x^6 + \ldots\right\}$$

The general solution of the d.e. is, of course, then obtained by adding this series to the one in 20A.

FRAME 22

In Example 1 the values of c were 0 and $\frac{1}{2}$, in Example 2 they were 1 and $\frac{1}{2}$ and in Example 3, $+\frac{1}{3}$ and $-\frac{1}{3}$. In none of these cases did the values of c differ by an integer. When this happens, the method of solution is affected in one of two ways, as will now be demonstrated by suitable examples.

Example 4   Solve the equation

$$y'' + xy' + y = 0$$

You can begin this in the usual way by obtaining the indicial equation and solving it.

***

22A

$$a_0 c(c-1) = 0$$

$$c = 0 \text{ or } 1$$

FRAME 23

You will notice that the roots of the indicial equation differ by an integer, 1.

We shall now proceed in the usual way and see what happens.

The indicial equation was obtained by equating the coefficient of x^{c-2} to zero, so we turn our attention next to the coefficient of x^{c-1}, which you should now write down.

****

23A

$$a_1(c+1)c$$

FRAME 24

Equating the coefficient of x^{c-1} to zero gives

$$a_1(c+1)c = 0 \qquad (24.1)$$

You will notice that this is similar to equation (17.1) in Example 3, in that only a_1 occurs, not a_1 and a_0. However, it differs in that we cannot say that $(c+1)c$ is not zero for either value of c, so it cannot be said that $a_1 = 0$ for both values of c.

When $c = 1$ we have $2a_1 = 0$ and therefore $a_1 = 0$

But when $c = 0$, equation (24.1) is satisfied whatever the value of a_1 so that the value of a_1 is left undetermined.

FRAME 25

Only the y'' term contributed to the coefficients of x^{c-2} and x^{c-1}, but from x^c onwards contributions are made by all three terms on the L.H.S. of the d.e., i.e. by y'', xy' and y. Therefore the coefficients of x^c, x^{c+1}, x^{c+2} etc. will all follow the same pattern and all 'a's from a_2 onwards will be given by a recurrence relation.

SOLUTION IN SERIES

FRAME 25 continued

Now obtain this recurrence relation by equating to zero the coefficient of x^{c+r} in $y'' + xy' + y$.

****

25A

$$a_{r+2} = -\frac{a_r}{c+r+2} \quad \text{for} \quad r \geq 0$$

FRAME 26

Taking first $c = 0$, we have already seen that in this case the value of a_1 remains undetermined. The recurrence relation now becomes

$$a_{r+2} = -\frac{a_r}{r+2}$$

Write down the results obtained by substituting $r = 0, 1, 2, 3$ in this relation.

****

26A

$r = 0$ $a_2 = -\dfrac{a_o}{2}$

$r = 1$ $a_3 = -\dfrac{a_1}{3}$

$r = 2$ $a_4 = -\dfrac{a_2}{4}$

$r = 3$ $a_5 = -\dfrac{a_3}{5}$

FRAME 27

The series deriving from $c = 0$ is therefore

$$x^o \left(a_o + a_1 x - \frac{a_o}{2} x^2 - \frac{a_1}{3} x^3 + \frac{a_o}{2.4} x^4 + \frac{a_1}{3.5} x^5 - \ldots \right)$$

i.e. $a_o \left(1 - \dfrac{x^2}{2} + \dfrac{x^4}{2.4} - \ldots \right) + a_1 \left(x - \dfrac{x^3}{3} + \dfrac{x^5}{3.5} \ldots \right)$

FRAME 27 continued

As a_0 and a_1 are undetermined, we can represent them by arbitrary constants A and B, so we have the solution

$$A\left(1 - \frac{x^2}{2} + \frac{x^4}{2.4} - \ldots\right) + B\left(x - \frac{x^3}{3} + \frac{x^5}{3.5} \ldots\right)$$

The present situation is rather puzzling, because this solution already has the two arbitrary constants required in the general solution. It looks as if we shall have too many arbitrary constants when we add the solution corresponding to $c = 1$.

Well, let's see what happens when $c = 1$. It was concluded in FRAME 24 that in this case $a_1 = 0$. The next step is to write down the recurrence relation and deduce expressions for a_2, a_3, a_4, a_5, and we suggest you do this.

****

27A

$$a_{r+2} = -\frac{a_r}{r+3}$$

$$a_2 = -\frac{a_0}{3}$$

$$a_3 = 0 \quad \text{as} \quad a_1 = 0$$

$$a_4 = -\frac{a_2}{5}$$

$$a_5 = 0$$

FRAME 28

The series corresponding to $c = 1$ is therefore

$$x\left(a_0 - \frac{a_0}{3}x^2 + \frac{a_0}{3.5}x^4 - \ldots\right)$$

i.e. $a_0\left(x - \frac{x^3}{3} + \frac{x^5}{3.5} - \ldots\right)$

But this series already forms part of the solution obtained in FRAME 27 from $c = 0$, so it has nothing new to contribute. We see, therefore, that the solution given by the lower value of c was, in fact, the general solution

SOLUTION IN SERIES

FRAME 28 continued

$$y = A\left(1 - \frac{x^2}{2} + \frac{x^4}{2.4} - \ldots\right) + B\left(x - \frac{x^3}{3} + \frac{x^5}{3.5} - \ldots\right)$$

FRAME 29

In Example 4, the method of obtaining the general solution of the d.e. was broadly the same as it was in the previous examples. However, this is not always the case when the roots of the indicial equation differ by an integer.

Bessel's equation of order n was referred to in the introduction, and in Example 3 it was solved for $n = \frac{1}{3}$. We shall now see what happens when n is an integer.

The equation is $\quad x^2 y'' + xy' + (x^2 - n^2)y = 0$

Starting with the usual trial solution, obtain the indicial equation.

****

29A

Coefficient of $x^c = (c^2 - n^2)a_0$

Indicial equation: $\quad c^2 - n^2 = 0$

FRAME 30

The roots of the indicial equation are $c = \pm n$, and if n is an integer the difference between the roots will also be an integer.

Now equate the coefficient of x^{c+1} to zero and deduce the value of a_1.

****

30A

$a_1(c + 1)c + a_1(c + 1) - n^2 a_1 = 0$

i.e. $\quad \{(c + 1)^2 - n^2\}a_1 = 0 \qquad (30A.1)$

As neither value of c makes $\{(c + 1)^2 - n^2\}$ zero, it can be concluded that
$$a_1 = 0$$

FRAME 31

The term $x^2 y$ on the L.H.S. of the d.e. has not contributed to the coefficients of x^c and x^{c+1}, but from x^{c+2} onwards this will cease to be the case. Hence equating the coefficient of x^{c+r} to zero will give a recurrence relation valid for $r \geq 2$. Find this recurrence relation (giving a_r in terms of a_{r-2}), and deduce the values of a_3, a_5, a_7 etc. for both values of c.

****

31A

$$a_r = -\frac{a_{r-2}}{(c+r)^2 - n^2} \qquad (31A.1)$$

$a_3 = a_5 = a_7 = \ldots = 0$ for either value of c.

FRAME 32

Assuming, for the sake of argument, that n is a <u>positive</u> integer, find the first four terms in the solution corresponding to $c = n$.

****

32A

$$a_r = -\frac{a_{r-2}}{r(2n+r)}$$

$$\therefore \quad a_2 = -\frac{a_0}{2(2n+2)} = -\frac{a_0}{2^2 \cdot 1(n+1)}$$

$$a_4 = -\frac{a_2}{4(2n+4)} = -\frac{a_2}{2^2 \cdot 2(n+2)}$$

$$a_6 = -\frac{a_4}{6(2n+6)} = -\frac{a_4}{2^2 \cdot 3(n+3)}$$

Denoting a_0 by an arbitrary constant A, the solution is

$$A\left\{x^n - \frac{x^{n+2}}{2^2 1!(n+1)} + \frac{x^{n+4}}{2^4 2!(n+2)(n+1)} - \frac{x^{n+6}}{2^6 3!(n+3)(n+2)(n+1)} + \ldots\right\}$$

If $A = \dfrac{1}{2^n n!}$ the result is "Bessel's function of order n of the first kind" and is denoted by $J_n(x)$.

SOLUTION IN SERIES

FRAME 33

Turning next to $c = -n$, the recurrence relation then becomes

$$a_r = -\frac{a_{r-2}}{r(r - 2n)}$$

so that we have

$$a_2 = -\frac{a_o}{2(2 - 2n)} = -\frac{a_o}{2^2 \cdot 1(1 - n)}$$

$$a_4 = -\frac{a_2}{4(4 - 2n)} = -\frac{a_2}{2^2 \cdot 2(2 - n)}$$

$$a_6 = -\frac{a_4}{6(6 - 2n)} = -\frac{a_4}{2^2 \cdot 3(3 - n)}$$

By considering in turn $n = 1$, $n = 2$, $n = 3$, can you see why $c = -n$ will not give a solution?

****

33A

If $n = 1$, a_2 becomes infinite.

" $n = 2$, a_4 "   "

" $n = 3$, a_6 "   "

By this time it should be obvious that n being a positive integer makes an 'a' infinite, so a series solution is not obtained by putting $c = -n$.

FRAME 34

In this type of situation, then, the usual procedure only gives one series solution (the one in 32A) which we know cannot be the general solution, having only one arbitrary constant. It is possible to find a second series solution, with the necessary second arbitrary constant, but the explanation of the method and the working involved are rather lengthy and we feel that most technologists will not find a detailed study of this particular problem sufficiently rewarding to justify the time spent. For those who are interested, a worked example is given in APPENDIX B.

FRAME 35

Mention must also be made of the case when the roots of the indicial equation are equal. This would happen, for instance, with Bessel's equation of order zero, where the indicial equation is $c^2 = 0$. Having only one value of c, we shall only get one series solution, with one arbitrary constant, using the method described in this programme. The remarks made in the previous frame apply equally to this case, and you can see a worked example in APPENDIX A, if you wish.

FRAME 36

Legendre's Equation

As has already been mentioned, Bessel's equation occurs in certain physical problems. It can arise after separation of the variables in a particular partial differential equation (Laplace's Equation). When a different system of coordinates is used, we get instead a d.e. of the form

$$(1 - x^2)y'' - 2xy' + n(n + 1)y = 0$$

This is known as Legendre's equation of order n.

Let us now solve Legendre's equation of order 1, i.e.

$$(1 - x^2)y'' - 2xy' + 2y = 0$$

Make a start by writing down and solving the indicial equation.

****

36A

Coefficient of $x^{c-2} = a_0 c(c - 1)$

Equating this to zero gives $c = 0$ *or* 1.

FRAME 37

Now equate the coefficient of x^{c-1} to zero and state what conclusions can be drawn about the value of a_1.

****

SOLUTION IN SERIES

37A

$$a_1(c+1)c = 0$$

If $c = 0$, a_1 can take any value i.e. it is undetermined

If $c = 1$, $a_1 = 0$.

FRAME 38

Now go ahead and find the values of a_2, a_3, a_4, a_5, a_6 for the case when $c = 0$, and hence write down the series solution corresponding to this value of c.

****

38A

$$a_2 = -a_0$$
$$a_3 = 0 \quad \text{and} \quad \therefore \; a_5 = 0$$
$$a_4 = \frac{1}{3} a_2$$
$$a_6 = \frac{3}{5} a_4$$

Series solution is $x^0(a_0 + a_1 x - a_0 x^2 - \frac{1}{3} a_0 x^4 - \frac{1}{3} \cdot \frac{3}{5} a_0 x^6 - \ldots)$

Denoting a_0 and a_1 by arbitrary constants A and B respectively this becomes

$$A\left(1 - x^2 - \frac{1}{3} x^4 - \frac{1}{5} x^6 - \ldots\right) + Bx$$

FRAME 39

You will notice that this is similar to Example 4, in that the solution given by the smaller value of c has the two arbitrary constants necessary for the general solution. You should now check that the solution given by $c = 1$ is already included in the solution just found.

****

6:42 SOLUTION IN SERIES

39A

If $c = 1$, $a_1 = 0$

$a_2 = 0$

$\therefore$ $a_3 = a_4 = a_5 = a_6 = \ldots = 0$

This leaves only $a_0 x$, which is already covered by the Bx in the solution obtained from $c = 0$.

FRAME 40

The complete solution, then, of Legendre's equation of order 1 is

$$y = A\left(1 - x^2 - \frac{1}{3}x^4 - \frac{1}{5}x^6 - \ldots\right) + Bx \qquad (40.1)$$

You have seen that, as in Example 4 where the roots differed by an integer, the complete solution is given by the lower value of c.

The solution (40.1) differs from those obtained in all the previous examples in this programme in that it does not consist of two infinite series, but of one infinite series and a series which terminates after a limited number of terms (in this example, only one). This always happens with Legendre's equation when n is a positive integer. The terminating series is then a polynomial of degree n, and, if the arbitrary constant in each case is chosen so that the value of the polynomial is 1 when $x = 1$, the functions known as Legendre's polynomials are obtained. The notation $P_n(x)$ is used. Thus $P_1(x) = x$.

FRAME 41

$\sum$ notation

You are probably aware of the $\sum$ notation which can be used for the sum of a number of terms all of the same kind. Its advantage is that it is more compact. For example, instead of the trial solution being written as

$$y = x^c(a_0 + a_1 x + a_2 x^2 + \ldots + a_r x^r + \ldots)$$

SOLUTION IN SERIES

FRAME 41 continued

it could be expressed as

$$y = x^c \sum_{r=0}^{\infty} a_r x^r \quad \text{or} \quad \sum_{r=0}^{\infty} a_r x^{c+r}.$$

This notation can be used throughout the whole of the working in solution in series problems and the rather cumbersome lay-outs incurred in writing out y, y', y" at length are then avoided. However, the non-mathematician is not usually sufficiently familiar with working with $\sum$ notation to be really happy with it, so it has not been used in this programme. For those who are interested, an example demonstrating its use is given in APPENDIX C.

FRAME 42

Miscellaneous Examples

In this frame some miscellaneous examples are given for you to try. Answers are supplied in FRAME 43 and hints have been provided in some cases.

Use a trial solution of the form (2.1) to solve the following d.e.'s.

1. $3xy'' + (1 - 3x)y' - 3y = 0$

2. $2x^2 y'' - xy' + (1 - x^2)y = 0$

3. $xy'' + 2y' + xy = 0$

4. $(x - x^2)y'' - 3y' + 2y = 0$

FRAME 43

1. $y = A\left(1 + 3x + \frac{3^2 x^2}{4} + \frac{3^3 x^3}{4 \cdot 7} + \ldots\right)$

    $+ Bx^{2/3}\left(1 + x + \frac{x^2}{2!} + \frac{x^3}{3!} + \ldots\right)$

2. $y = Ax^{\frac{1}{2}}\left(1 + \frac{x^2}{3.2} + \frac{x^4}{3.7.2.4} + \frac{x^6}{3.7.11.2.4.6} + \ldots\right)$

$+ Bx\left(1 + \frac{x^2}{2.5} + \frac{x^4}{2.4.5.9} + \frac{x^6}{2.4.6.5.9.13} + \ldots\right)$

3. $c = -1$ or 0

   $c = -1$ gives the solution

   $$y = Ax^{-1}\left(1 - \frac{x^2}{2!} + \frac{x^4}{4!} - \ldots\right) + B\left(1 - \frac{x^2}{3!} + \frac{x^4}{5!} - \ldots\right)$$

   $c = 0$ repeats the second series in this solution.

4. $c = 0$ or 4

   $c = 0$ gives the solution

   $$y = A\left(1 + \frac{2x}{3} + \frac{x^2}{3}\right) + Bx^4\left(1 + 2x + 3x^2 + 4x^3 + \ldots\right)$$

   You may have thought that, because $a_3 = 0$, $a_4 = a_5 = a_6 = \ldots$
   $\ldots = 0$ too, but $a_4 = \frac{c+1}{c} a_3$, so if $a_3 = 0$ and $c = 0$, a_4 is indeterminate.

   Or think of it another way: $ca_4 = (c+1)a_3$
   If $c = 0$ and $a_3 = 0$, this relationship is satisfied whatever the value of a_4. The second arbitrary constant B is assigned to a_4.

   $c = 4$ repeats the infinite series.

SOLUTION IN SERIES

APPENDIX A

Roots of Indicial Equation equal

The method of solution in this case will be shown by considering Bessel's equation of order zero, i.e.

$$xy'' + y' + xy = 0$$

In FRAME 29 you were asked to write down the indicial equation for Bessel's equation of order n. Putting n = 0 in the result gives

$$c^2 = 0$$

whose roots are both equal to zero.

Putting n = 0 in the other results obtained for Bessel's equation of order n, the equation (30A.1) becomes $(c + 1)^2 a_1 = 0$

giving $a_1 = 0$ as for other values of n, and the recurrence relation (31A.1) becomes

$$a_r = -\frac{a_{r-2}}{(c + r)^2}.$$

Hence $a_3 = a_5 = a_7 = \ldots = 0$

and $a_2 = -\dfrac{a_0}{(c + 2)^2}$, $a_4 = -\dfrac{a_2}{(c + 4)^2}$, $a_6 = -\dfrac{a_4}{(c + 6)^2}$ etc.

Let us write $z = a_0 x^c \left\{ 1 - \dfrac{1}{(c + 2)^2} x^2 + \dfrac{1}{(c + 2)^2 (c + 4)^2} x^4 - \dfrac{1}{(c + 2)^2 (c + 4)^2 (c + 6)^2} x^6 \ldots \right\}$

This is the series we get when we substitute the expressions obtained above for a_1, a_2, a_3 etc. and it will be a solution of the d.e. when $c = 0$.

If the series denoted by z is substituted for y in the L.H.S. of the d.e., all the terms cancel out except one, $a_0 c^2 x^{c-1}$. (If you wish to verify this for yourself you are advised to use the kind of lay-out shown on Page 6:30.)
Putting this another way,

$$xz'' + z' + xz = a_0 c^2 x^{c-1} \qquad (A.1)$$

and, of course, you will see that $c = 0$ makes the R.H.S. zero and hence makes z in that case a solution of the original d.e.

APPENDIX A continued

In what follows next, $\frac{\partial}{\partial m}(x^m)$ will be required and this will now be obtained in case you are not familiar with the result.

$$\text{If} \quad f = x^m$$
$$\log_e f = m \log_e x$$
$$\frac{1}{f} \frac{\partial f}{\partial m} = \log_e x$$
$$\frac{\partial f}{\partial m} = f \log_e x$$
$$\text{i.e.} \quad \frac{\partial}{\partial m}(x^m) = x^m \log_e x$$

If both sides of equation (A.1) are differentiated partially with respect to c, we get

$$x \frac{\partial}{\partial c} z'' + \frac{\partial}{\partial c} z' + x \frac{\partial}{\partial c} z = a_o (2cx^{c-1} + c^2 x^{c-1} \log_e x) \qquad (A.2)$$

Now, as you know, differentiating z' with respect to c gives the same result as differentiating $\frac{\partial z}{\partial c}$ with respect to x. That is to say, it makes no difference whether we differentiate with respect to x before differentiating with respect to c, or vice versa.

∴ Equation (A.2) is equivalent to

$$x \left(\frac{\partial z}{\partial c}\right)'' + \left(\frac{\partial z}{\partial c}\right)' + x \frac{\partial z}{\partial c} = a_o c (2x^{c-1} + cx^{c-1} \log_e x)$$

Now, putting $c = 0$ will make the R.H.S. zero, just as it did in equation (A.1). Hence, when $c = 0$, $\frac{\partial z}{\partial c}$, as well as z, is a solution of $xy'' + y' + xy = 0$ and this gives us the second solution which we require.

The next step is to find $\frac{\partial z}{\partial c}$, and if you refer back to the series which represents z, you may think the differentiation appears rather formidable.

It will help to make the working clearer if we write

$$z = a_o x^c V \quad \text{where} \quad V = 1 - \frac{1}{(c+2)^2} x^4 + \frac{1}{(c+2)^2 (c+4)^2} x^4 - \frac{1}{(c+2)^2 (c+4)^2 (c+6)^2} x^6 + \ldots$$

SOLUTION IN SERIES

APPENDIX A continued

Then $\dfrac{\partial z}{\partial c} = a_o (Vx^c \log_e x + x^c \dfrac{\partial V}{\partial c})$

To find $\dfrac{\partial V}{\partial c}$, it will be necessary to differentiate terms like

$\dfrac{1}{(c+2)^2(c+4)^2}$ and $\dfrac{1}{(c+2)^2(c+4)^2(c+6)^2}$ with respect to c. If we can find out what happens in the case $\dfrac{1}{(c+2)^2(c+4)^2}$, perhaps we shall be able to deduce the corresponding results for the more difficult terms.

$$\text{Let} \quad w = \dfrac{1}{(c+2)^2(c+4)^2}$$

$$\log_e w = -2\log_e(c+2) - 2\log_e(c+4)$$

$$\dfrac{1}{w}\dfrac{dw}{dc} = -\dfrac{2}{c+2} - \dfrac{2}{c+4}$$

$$\dfrac{dw}{dc} = -\dfrac{2}{(c+2)^2(c+4)^2}\left\{\dfrac{1}{c+2} + \dfrac{1}{c+4}\right\}$$

From this it can be seen that, similarly, differentiating

$\dfrac{1}{(c+2)^2(c+4)^2(c+6)^2}$ with respect to c will yield

$$-\dfrac{2}{(c+2)^2(c+4)^2(c+6)^2}\left\{\dfrac{1}{c+2} + \dfrac{1}{c+4} + \dfrac{1}{c+6}\right\}.$$

Returning to $\dfrac{\partial z}{\partial c}$, we have

$$\dfrac{\partial z}{\partial c} = a_o x^c V \log_e x + a_o x^c \dfrac{\partial V}{\partial c}$$

$$= z \log_e x + a_o x^c \left\{ \dfrac{2}{(c+2)^3} x^2 - 2 \dfrac{\dfrac{1}{c+2} + \dfrac{1}{c+4}}{(c+2)^2(c+4)^2} x^4 \right.$$

$$\left. + 2 \dfrac{\dfrac{1}{c+2} + \dfrac{1}{c+4} + \dfrac{1}{c+6}}{(c+2)^2(c+4)^2(c+6)^2} x^6 \ldots \right\}$$

We now have the two series, z and $\dfrac{\partial z}{\partial c}$, which are solutions of the d.e. when $c = 0$.

APPENDIX A continued

Putting $c = 0$ and denoting a_o by an arbitrary constant A, z becomes

$$A\left(1 - \frac{1}{2^2}x^2 + \frac{1}{2^2.4^2}x^4 - \frac{1}{2^2.4^2.6^2}x^6 + \ldots\right)$$

Putting $c = 0$ and denoting a_o by an arbitrary constant B, $\frac{\partial z}{\partial c}$ becomes

$$B\left(1 - \frac{1}{2^2}x^2 + \frac{1}{2^2.4^2}x^4 - \frac{1}{2^2.4^2.6^2}x^6 + \ldots\right)\log_e x$$

$$+ B\left(\frac{1}{2^2}x^2 - \frac{1+\frac{1}{2}}{2^2.4^2}x^4 + \frac{1+\frac{1}{2}+\frac{1}{3}}{2^2.4^2.6^2}x^6 - \ldots\right)$$

The general solution is

$$y = (A + B\log_e x)\left(1 - \frac{1}{2^2}x^2 + \frac{1}{2^2.4^2}x^4 - \frac{1}{2^2.4^2.6^2}x^6 + \ldots\right)$$

$$+ B\left(\frac{1}{2^2}x^2 - \frac{1+\frac{1}{2}}{2^2.4^2}x^4 + \frac{1+\frac{1}{2}+\frac{1}{3}}{2^2.4^2.6^2}x^6 - \ldots\right)$$

SOLUTION IN SERIES   6:49

APPENDIX B

If you have not read APPENDIX A, you should do so before proceeding.

Roots of Indicial Equation differing by an integer and making an 'a' infinite.

The method of solution in this case will be shown by considering Bessel's equation of order one i.e.

$$x^2 y'' + xy' + (x^2 - 1)y = 0$$

In FRAME 29 you were asked to write down the indicial equation for Bessel's equation of order n. Putting $n = 1$ in the result gives

$$c^2 - 1 = 0$$

whose roots are -1 and $+1$, i.e. differing by 2.

Putting $n = 1$ in the other results obtained for Bessel's equation of order n, the equation (30A.1) becomes $\{(c+1)^2 - 1\}a_1 = 0$

giving $a_1 = 0$ as for other values of n,

and the recurrence relation (31A.1) becomes $a_r = -\dfrac{a_{r-2}}{(c+r)^2 - 1}$.

Hence $a_3 = a_5 = a_7 = \ldots = 0$

and $a_2 = -\dfrac{a_0}{(c+1)(c+3)}$, $a_4 = -\dfrac{a_2}{(c+3)(c+5)}$, $a_6 = -\dfrac{a_4}{(c+5)(c+7)}$

etc.

Let us write

$$z = a_0 x^c \left\{ 1 - \frac{1}{(c+1)(c+3)} x^2 + \frac{1}{(c+1)(c+3)^2(c+5)} x^4 - \frac{1}{(c+1)(c+3)^2(c+5)^2(c+7)} x^6 + \ldots \right\}$$

This is the series we get when we substitute the expressions obtained above for a_1, a_2, a_3 etc. When $c = 1$, the solution

$$Ax \left\{ 1 - \frac{1}{2 \cdot 4} x^2 + \frac{1}{2 \cdot 4^2 \cdot 6} x^4 - \frac{1}{2 \cdot 4^2 \cdot 6^2 \cdot 8} x^6 + \ldots \right\}$$

is obtained, but if we substitute $c = -1$, the other root of the indicial equation, the coefficients become infinite because of the factor $(c + 1)$ in the denominators.

APPENDIX B continued

This difficulty can be overcome by replacing a_o by $(c + 1)k$, and replacing the previous condition $a_o \neq 0$ by the condition $k \neq 0$. We then have

$$z = kx^c \left\{ (c + 1) - \frac{1}{c + 3} x^2 + \frac{1}{(c + 3)^2(c + 5)} x^4 - \frac{1}{(c + 3)^2(c + 5)^2(c + 7)} x^6 + \ldots \right\} \qquad (B.1)$$

If this series is substituted for y in the L.H.S. of the d.e., all the terms cancel out except one, $kx^c(c + 1)^2(c - 1)$. As in APPENDIX A, the working for this is not set out here - it is left to you to verify the result if you wish, and again the lay-out as shown on Page 6:30 is recommended.

We have, then,

$$x^2 z'' + xz' + (x^2 - 1)z = kx^c(c + 1)^2(c - 1)$$

The situation is now similar to that arrived at when equation (A.1) was written down in APPENDIX A. There was then a factor c^2 on the R.H.S. and, because of that, $\frac{\partial z}{\partial c}$, as well as z, was a solution of the d.e. when $c = 0$. In the present case, the squared factor $(c + 1)^2$ on the R.H.S. makes $\frac{\partial z}{\partial c}$, as well as z, a solution of the d.e. when $c = -1$.

$\frac{\partial z}{\partial c}$ is found in much the same way as in APPENDIX A - this time we shall just state the result.

$$\frac{\partial z}{\partial c} = z \log_e x + kx^c \left\{ 1 + \frac{1}{(c + 3)^2} x^2 - \frac{\frac{2}{c + 3} + \frac{1}{c + 5}}{(c + 3)^2(c + 5)} x^4 + \frac{\frac{2}{c + 3} + \frac{2}{c + 5} + \frac{1}{c + 7}}{(c + 3)^2(c + 5)^2(c + 7)} x^6 - \ldots \right\}$$

We have already found one solution, corresponding to $c = 1$, so if we are going to get two further solutions from $c = -1$, it would appear that we are going to have three solutions altogether, giving a general solution with three arbitrary constants whereas for a second order d.e. there should be only two. However, substituting $c = -1$ in (B.1) gives the solution

SOLUTION IN SERIES   6:51

<u>APPENDIX B</u> continued

$$Bx^{-1}\left(-\frac{1}{2}x^2 + \frac{1}{2^2.4}x^4 - \frac{1}{2^2.4^2.6}x^6 + \ldots\right)$$

and this can be written

$$-\frac{B}{2}x\left(1 - \frac{1}{2.4}x^2 + \frac{1}{2.4^2.6}x^4 - \ldots\right)$$

which is essentially the same as the solution obtained using $c = 1$. So there are only two linearly independent solutions after all. The second one, obtained by putting $c = -1$ in $\frac{\partial z}{\partial c}$, is

$$Bx\log_e x\left(1 - \frac{1}{2.4}x^2 + \frac{1}{2.4^2.6}x^4 - \frac{1}{2.4^2.6^2.8}x^6 + \ldots\right)$$

$$+ Bx^{-1}\left(1 + \frac{1}{2^2}x^2 - \frac{1+\frac{1}{4}}{2^2.4}x^4 + \frac{1+\frac{1}{2}+\frac{1}{6}}{2^2.4^2.6}x^6 - \ldots\right)$$

The general solution is

$$y = (A + B\log_e x)x\left(1 - \frac{1}{2.4}x^2 + \frac{1}{2.4^2.6}x^4 - \frac{1}{2.4^2.6^2.8}x^6 + \ldots\right)$$

$$+ Bx^{-1}\left(1 + \frac{1}{2^2}x^2 - \frac{1+\frac{1}{4}}{2^2.4}x^4 + \frac{1+\frac{1}{2}+\frac{1}{6}}{2^2.4^2.6}x^6 - \ldots\right)$$

APPENDIX C

An example using Σ notation

The equation which has been chosen to show the use of Σ notation is

$$2x^2y'' - xy' + (1 - 2x)y = 0$$

This was solved in Example 2 in FRAMES 13 - 14.

The trial solution is
$$y = \sum_{r=0}^{\infty} a_r x^{c+r}$$

so
$$y' = \sum_{r=0}^{\infty} a_r (c + r) x^{c+r-1}$$

and
$$y'' = \sum_{r=0}^{\infty} a_r (c + r)(c + r - 1) x^{c+r-2}$$

Substituting these expressions in $2x^2y'' - xy' + (1 - 2x)y$ leads to

$$2\sum_{r=0}^{\infty} a_r(c+r)(c+r-1)x^{c+r} - \sum_{r=0}^{\infty} a_r(c+r)x^{c+r} + \sum_{r=0}^{\infty} a_r x^{c+r} - 2\sum_{r=0}^{\infty} a_r x^{c+r+1}$$

On combining the first three $\sum$'s, which all involve x^{c+r}, this becomes

$$\sum_{r=0}^{\infty} \{2(c+r)(c+r-1) - (c+r-1)\} a_r x^{c+r} - 2\sum_{r=0}^{\infty} a_r x^{c+r+1}$$

i.e.
$$\sum_{r=0}^{\infty} (2c + 2r - 1)(c + r - 1) a_r x^{c+r} - 2\sum_{r=0}^{\infty} a_r x^{c+r+1}$$

Now x^{c+r+1} when $r = 0$ is x^{c+r} when $r = 1$, so $\sum_{r=0}^{\infty} a_r x^{c+r+1}$ can be rewritten as $\sum_{r=1}^{\infty} a_{r-1} x^{c+r}$. This is done to bring all powers of x in the $\sum$'s down to the lowest one present (x^{c+r} in this case).

The next step is to equate to zero the various powers of x in

$$\sum_{r=0}^{\infty} (2c + 2r - 1)(c + r - 1) a_r x^{c+r} - 2\sum_{r=1}^{\infty} a_{r-1} x^{c+r}$$

When $r = 0$, only the first $\sum$ makes a contribution and gives the term

SOLUTION IN SERIES   6:53

APPENDIX C continued

$(2c - 1)(c - 1)a_o x^c$. Equating this coefficient to zero, we get

$$(2c - 1)(c - 1) = 0 \quad \text{as} \quad a_o \neq 0.$$

This is, of course, the indicial equation, as obtained in answer frame 13A.
From $r = 1$ onwards, both $\sum$'s make a contribution.

$\therefore$ for $r \geq 1$, coefficient of $x^{c+r} = (2c + 2r - 1)(c + r - 1)a_r - 2a_{r-1}$
Equating this to zero gives the recurrence relation

$$a_r = \frac{2a_{r-1}}{(2c + 2r - 1)(c + r - 1)}$$

as obtained in answer frame 14A.

The roots of the indicial equation are $\frac{1}{2}$ and 1.

Taking first $c = 1$, the recurrence relation becomes

$$a_r = \frac{2a_{r-1}}{(2r + 1)r}$$

Repeated application of this formula leads to

$$a_r = \frac{2^r}{r! \, 3.5.7.....(2r + 1)} a_o = \frac{2^r \, 2.4.6.....2r}{r!(2r + 1)!} a_o$$

$$= \frac{2^r . 2^r \, r!}{r!(2r + 1)!} a_o = \frac{2^{2r}}{(2r + 1)!} a_o$$

The solution given by $c = 1$ can therefore be written in the form

$$Ax\left\{1 + \sum_{r=1}^{\infty} \frac{2^{2r}}{(2r + 1)!} x^r\right\}$$

where A is the arbitrary constant assigned to a_o.
(This is solution (14A.1) written in $\sum$ form.)

When $c = \frac{1}{2}$, the recurrence relation becomes

$$a_r = \frac{2a_{r-1}}{2r(r - \frac{1}{2})}$$

i.e. $$a_r = \frac{2a_{r-1}}{r(2r - 1)}$$

APPENDIX C continued

By applying this formula repeatedly, and simplifying the result, we obtain

$$a_r = \frac{2^{2r-1}}{r(2r-1)!} a_o$$

Denoting a_o by an arbitrary constant B, we then have the solution

$$Bx^{\frac{1}{2}}\left\{1 + \sum_{r=1}^{\infty} \frac{2^{2r-1}}{r(2r-1)!} x^r\right\}$$

which is, of course, (14A.2) written in $\sum$ form.

The complete solution is

$$y = Ax\left\{1 + \sum_{r=1}^{\infty} \frac{2^{2r}}{(2r+1)!} x^r\right\} + Bx^{\frac{1}{2}}\left\{1 + \sum_{r=1}^{\infty} \frac{2^{2r-1}}{r(2r-1)!} x^r\right\}$$